W9-CGY-419

FARADAY TO EINSTEIN: CONSTRUCTING MEANING IN SCIENTIFIC THEORIES

SCIENCE AND PHILOSOPHY

This series has been established as a forum for contemporary analysis of philosophical problems which arise in connection with the construction of theories in the physical and the biological sciences. Contributions will not place particular emphasis on any one school of philosophical thought. However, they will reflect the belief that the philosophy of science must be firmly rooted in an examination of actual scientific practice. Thus, the volumes in this series will include or depend significantly upon an analysis of the history of science, recent or past. The Editors welcome contributions from scientists as well as from philosophers and historians of science.

NANCY J. NERSESSIAN

Faraday to Einstein: Constructing Meaning in Scientific Theories

1984 **MARTINUS NIJHOFF PUBLISHERS**
a member of the KLUWER ACADEMIC PUBLISHERS GROUP
DORDRECHT / BOSTON / LANCASTER

Distributors

for the United States and Canada: Kluwer Academic Publishers, 190 Old Derby Street, Hingham, MA 02043, USA
for the UK and Ireland: Kluwer Academic Publishers, MTP Press Limited, Falcon House, Queen Square, Lancaster LA1 1RN, England
for all other countries: Kluwer Academic Publishers Group, Distribution Center, P.O. Box 322, 3300 AH Dordrecht, The Netherlands

Library of Congress Cataloging in Publication Data

Nersessian, Nancy J.
Faraday to Einstein: Constructing meaning in scientific theories.

(Science and Philosophy)
Includes index.
1. Science -- Philosophy. I. Title. II. Series.
Q175.N3885 1984 501 84-14679

ISBN 90-247-2997-1 (this volume)

Copyright

PRINTED IN THE NETHERLANDS

TO THE MEMORY OF MY PARENTS

Edward Simon Nersessian
Marjorie Janice Stevenson

Table of contents

Preface

Einstein often expressed the sentiment that "the eternal mystery of the world is its comprehensibility," and that science is the means through which we comprehend it. However, nearly everyone – including scientists – agrees that the concepts of modern physics are quite *in*comprehensible: They are both unintelligible to the educated lay-person and to the scientific community itself, where there is much dispute over the interpretation of even (and especially) the most basic concepts. There is, of course, almost universal agreement that modern science quite adequately accounts for and predicts events, i.e., that its calculations work better than those of classical physics; yet the concepts of science are supposed to be descriptive of 'the world' as well – they should enable us to comprehend it. So, it is asked, and needs to be asked: Has modern physics failed in an important respect?

It failed with me as a physics student. I came to physics, as with most naïve students, out of a desire to know what the world is *really* like; in particular, to understand Einstein's conception of it. I thought I had grasped the concepts in classical mechanics, but with electrodynamics confusion set in and only increased with relativity and quantum mechanics. At that point I began even to doubt whether I had really understood the basic concepts of classical mechanics. The mathematics was fun and interesting and I sensed that something fascinating was being said about the world, but I could not figure out *what*: What does it mean to call something 'a field'? What kind of thing is it in the world? What does it mean to say that something is 'a wave *and* a particle'? How can we possibly say that Newtonian mechanics is a 'special case' of relativity when the pictures they present of the world are supposed to be so different? The mathematical formulae could not supply an answer, nor could my textbooks and teachers. I was told that such questions are 'philosophical' (i.e., '*meta*physical') and, while interesting, more appropriately dealt with by philosophers. But philosophy seemed the antithesis of science to me, so I stuck to my science courses.

Through the double fortune of being forced to take a general philosophy course in order to get my degree and of its being taught as though it were an introduction to the philosophy of science, I thought I had found what I was looking for. Here it was legitimate to raise the questions which had so disturbed my physics teachers, and I decided to become a philosopher. As a graduate student in philosophy, I was again frustrated. I found that philosophers of science wanted to work within what they called 'the context of justification', i.e., science reconstructed in the way it *should* be if it is to be a logically consistent system of thought; whereas such things as the concrete histories and roles of the concepts in science they relegated to 'the context of discovery' – to work in this context was forbidden for philosophers. They called it a fallacy of logic ('the Genetic Fallacy') to confuse or conflate these two points of view. Thus, they tended to avoid investigations of real science – and especially its history – in favor of what they called 'rational reconstructions', i.e., artificial maps of the logical relations between concepts, or between theories, and the evidence for them. Having come from science, the reason for the distinction was never really convincing to me: Since it is not possible to provide a 'recipe' for creating new concepts and theories, i.e., to provide a 'logic of discovery', it cannot be guaranteed that the actual process is rational, so that process must be irrelevant to understanding the nature of science and to distinguishing it from 'metaphysics' and 'pseudoscience'. But why not admit that there can be no *logic* of discovery, while at the same time admit that there are aspects of discovery that are essential to formulating an understanding of science? The more careful philosophers got with the distinction between contexts, the less they seemed to have anything to say about science as I knew it. Their questions about meaning, with their focus on the nature and necessities of language, seemed, at bottom, wholly unrelated to the problems which had led me from physics to philosophy.

I turned again to reading the works of scientists – only this time, rather than reading textbooks, I read the formative works in physics. Since my problems had begun with electrodynamics, I started with Faraday's *Experimental Researches* and found his struggle to conceptualize how forces could be transmitted across empty space responsive to the questions I had been asking: Not only that, I felt that I had begun to understand what a 'field' is. This feeling grew as I read Maxwell. What finally struck me is that the concepts of science are inventions – *made* by us – so their meanings can be better understood by examining their actual construction. Scientific concept formation is a process, involving the struggle to articulate what we mean when we say, e.g., that something is a 'field', and the key to understanding the concepts of science lies in examining the various phases of that process as it takes place in scientific practice. However, such an examination lies squarely in 'the context of discovery'. This conflict lead to a sort of 'intellectual schizophrenia': On the one hand I was involved in a discipline called 'the philosophy of science' and on the other I was reading the formative works of science, but I could not see how to put the two pictures together. Fortunately, this was not my problem alone. It was, and is, the problem of those post-positivistic philosophers of science who saw the inadequacy of that type of philosophical analysis for understanding *real* science. In order to understand real science, the dimension of discovery has to be added to philosophical analysis; in the case of meaning, a way has to be found to add the dimension of discovery to the notions of 'meaning' and 'meaning change' in science. This book is part of an ongoing struggle to do so.

Part I discusses the 'standard account' of meaning in scientific theories, and where it went wrong, and provides a succinct survey of the arguments leading to the so-called 'problem of incommensurability of meaning', showing that it presupposes a particular conception of meaning. We are led to the conclusion that the 'problem' should be seen simply as providing a refutation of that conception of meaning, indicating only that a new conception is needed. The inadequacies of the most popular alternative conception are discussed, concluding with a discussion of the method of analysis to be used in Parts II and III.

Part II discusses the formation of the present concept of 'electromagnetic field': from Faraday's 'lines of force' to Maxwell's 'Newtonian aether-field', to Lorentz' 'non-Newtonian aether-field', and finally to Einstein's 'independent reality', focusing on the beliefs and the problems which led to the various changes in what is meant by 'electromagnetic field'.

Part III draws the features of my approach together through an analysis of the process of concept formation as it appears in the case study and the introduction of the notion of a 'meaning schema' for scientific concepts. A 'meaning schema' for a particular concept contains within it key features of that concept at each phase of its development, with the reasoning process from one phase to the next supplying the connection between earlier and later forms of a concept. The construction of 'meaning schemata' for scientific concepts has the potential for making the concepts of science more comprehensible.

NIAS, Wassenaar
April 1984

NANCY J. NERSESSIAN

Acknowledgments

There are many people who have left their imprint on the pages that follow. I would like to acknowledge some of them here and to express my appreciation to them.

There are some general intellectual debts I will mention first. My turning to philosophy was the direct result of a course given by Milič Čapek, for which I remain grateful to him. Howard Stein's example led me to examine the works of scientists, and his support and criticism directed me; but I appreciate most that he always encouraged me to follow my own direction. From Joseph Agassi I have learned the difficult lesson of the value of criticism. Extensive discussions with James Hullett and William Berkson have played a significant role in shaping my ideas and in helping me to focus and articulate my views. I thank Floris Cohen for his careful reading of the penultimate version of this book and for his helpful comments.

I have also acquired debts of a more specific nature. I would like to thank A. J. Kox for providing me with some of the letters of Lorentz. I appreciate the comments made by H. B. G. Casimir on Chapter 5. I wish to thank the Museum Boerhaave, Leiden, for permission to quote from the Lorentz correspondence and for providing me with the photographs in Part II. I appreciate the efforts of the librarians at NIAS, Dick van den Kooij and Dinny Young, in acquiring books and inaccessible articles. I am grateful to Mark Wagner for his editorial suggestions and his assistance in proofreading, and to Corry Overdiep for her intelligent typing of the manuscript.

Finally, since even philosophers do not live on ideas alone, I want to express my gratitude for the generous financial support

I have received: to the Netherlands America Commission for Educational Exchange for appointing me as a Fulbright Scholar during 1981–1982; to the Netherlands Institute for Advanced Study in the Social Sciences and the Humanities, Wassenaar, for inviting me as a Fellow during 1983–1984; and to the Technical University of Twente for relieving me of university responsibilities during 1983–1984. I wish to thank NIAS, additionally, for providing the staff support and the ideal environment in which to complete this work.

PART I

The Philosophical Situation: A Critical Appraisal

> We must begin with the mistake and find out the truth in it.
> That is, we must uncover the source of the error; otherwise hearing what is true won't help us.
> It cannot penetrate when something is taking its place.
>
> Wittgenstein

Introduction

One of the central problems in 20th-century philosophy of science has been the nature of meaning in scientific theories. There is still no adequate response. In order to understand why this is so, I will survey the history of responses to the problem in as succinct a manner as possible, in an attempt to reveal what "mistake" is standing in the way of our seeing the "truth," before attacking the problem once again. This part of the book presents a critical overview of the most influential responses: the 'standard account' of logical positivism, the 'network view' of Duhem, Quine, and Feyerabend, the thesis of 'meaning variance' and the resulting 'problem of incommensurability', and the 'causal theory' of Kripke and Putnam.

The standard account, together with Carnap's modification of it, is presented in terms of its reductionist theory of meaning and its sharp dichotomies with their labyrinthine interconnections. The network view and its consequence, the problem of incommensurability of meaning, are traced to the successive collapse of these distinctions. The most widely accepted response to the incommensurability of meaning arises from the insight that the standard account of meaning in scientific theories is part of the more general Frege–Russell theory of meaning, in which meaning is viewed as having two aspects: sense and reference, with sense determining reference. What has come to be known as the 'causal theory' is the result of attempts to find a way to detach reference from sense; that is, to find a theory-independent way of accounting for reference. The causal theory is the most successful and most widely accepted response to the incommensurability of the network view. However, it does not provide a successful account of meaning (sense and reference) for scientific theories, i.e., its rigidity of reference and its essentialism do not square with scientific practice.

An examination of why this is so reveals the affinities of the causal theory with the standard account. They both make the same fundamental "mistake": They believe that an analysis of the necessities and nature of language alone is sufficient for the formulation of an adequate understanding of the nature of meaning in scientific theories. It is this belief that has prevented philosophers from providing an adequate theory of meaning for science. The 'linguistic turn' in philosophy has wrongly made responses to the problem of meaning in science parasitic on developments in the philosophy of language, and has thus led philosophy of science away from the subject of its study: science. The "truth" that has been obscured is simply this: Any theory of meaning adequate to science must be firmly rooted in an examination of scientific practices concerning meaning.

CHAPTER 1

The 'standard' account of meaning

1.1 General characterization

There are many variations of what has come to be known as the 'standard empiricist account of meaning in scientific theories' and thus any general characterization must, of necessity, be an oversimplification. However, it is possible to discuss the main theme without going into the details of the variations and still provide a useful analysis.[1] For our purpose all that is needed is a formulation of the main presuppositions of the account so that we may see that the retention of some of them is as necessary as the abandonment of others in order for 'meaning variance' and 'incommensurability' to pose serious problems. Before characterizing the account, some preliminary remarks need to be made. First, the account was intended as an account of empirical knowledge in general, with scientific knowledge taken as the paradigm. Although our concern is with science, some of the characterization will make reference to the more general issue. Second, we should keep in mind that the account and its 'reductionist theory of meaning' are part of the broader Frege–Russell theory of meaning. Finally, as Quine has so succinctly noted, for "old Vienna" meaning and verification are coextensive.[2] So, although our focus is meaning, to some extent the two notions will appear interwoven in the characterization.

The standard account has two main aspects: (1) a theory of meaning which involves 'reduction' of terms (later sentences) to empirical content, i.e., sensory experience, and (2) a 'double-language' conception of scientific theories, i.e., a theoretical language and an observation language. The account is perhaps best characterized by its position on the 'theoretical–observational' distinction: The observational and the theoretical components of the corpus of sentences which constitute empirical knowledge are sharply divided. The epistemological basis for this distinction is the parallel distinction between the 'given' and the 'conceptual' aspects of experience. Given experience is the untainted, pure experience of physical reality, while the conceptual aspect of experience involves human interpretation. The motivation behind making such a distinction is that if we are to have certain knowledge, we must have access to physical 'reality', independent of our interpretations.

Views on just how the two distinctions – one concerning language and the other concerning experience – are related underwent a series of modifications. Initially they were taken as strictly parallel. Theory construction is part of the conceptual aspect of experience; theoretical sentences are conceptual expressions. On the other hand, observation sentences directly express given experience, with no conceptual interpretation necessary. Theoretical sentences obtain their empirical meaning by 'reduction' to observation sentences. The insurmountable problem posed by having made such a sharp distinction is that of how to reconnect a theory with its observational basis.

The essential thesis of the standard account is that the observational situation provides whatever empirical meaning and verification empirical knowledge can have, for only in the observational situation do we have direct experience of physical reality: the given. It is the independent nature of the given which allows it to function as the connection between our theorizing and the external world. If observation sentences are direct expressions of given experience, then they are 'certain', 'indubitable', 'incorrigible', etc. reports of the empirical world. However, the major problem with this conception is that given experience is essentially private and subjective, while observation sentences are public and intersubjective. To avoid this problem, most empiricists quickly turned to a discussion of the observation sentences themselves (the 'linguistic turn'), merely claiming that they are direct expressions of given experience, whatever that might be. An accepted aura of vagueness descended upon the nature of the given, and 'certainty' gave way to 'maximal surity': The observational situation provides a maximally certain, theoretically unrevisable empirical 'core' which is the foundation upon which all empirical knowledge is secured. The attempt to make this thesis intelligible led to a hornet's nest of problems on which volumes have been written and which do not concern us here. Let me simply discuss what turned out to be the crucial challenge to considering the two distinctions – theoretical–observational and conceptual–given – as parallel. This discussion will, in turn, lead us to the 'double-language' view.

The essential criticism was posed independently by C. I. Lewis and R. Carnap: There can be no pre-conceptual ways of expressing knowledge claims in a language.[3] Carnap's formalization of this insight was most influential in codifying the double-language view and will be reserved for the next section. I will present the argument here as Lewis formulated it. First we need to introduce yet another distinction: the 'conceptual–theoretical'. What this distinction amounts to is that there is a difference *in kind* between a conceptual structure and a theory. A conceptual structure is definitive in nature, while a theory is factual. (We should keep in mind that this is a distinction concerning language – both the creation of a conceptual structure and of a theory are part of the conceptual aspect of experience.) Thus, the conceptual–theoretical distinction is based, in part, on the 'analytic–synthetic' distinction – a point I will return to in the following section. Given the conceptual–theoretical distinction, in order to maintain the distinction between the observation language and the theoretical language it is not necessary to maintain that observation sentences are not conceptual expressions. Not only is it not necessary, it is also not possible; this is the crux of the objection raised by Lewis and Carnap.

In *Mind and the World Order*, Lewis argued that experience of a pure given cannot be cognitive experience, and that knowledge cannot be constructed out of elements which directly mirror given experience, independent of any conceptual interpretation.[4] Any foundation for knowledge erected upon such experience would collapse into subjectivity, taking the whole edifice with it. Objectivity, and with it community of meaning and the possibility of verification and correction of knowledge claims, is not attained until whatever is given is interpreted.

Observation sentences are expressions of knowledge. Thus, even the most elementary sentences we can formulate, such as "Blue here now," are conceptually interpreted if they are meaningful. That is, we generate categories and concepts which are used to structure, classify, and understand the material given in sensuous experience. However, cognitive experience is not a two-staged process in which we first have experiences and then generate categories to interpret them; so we have no access to the 'pure given' and the 'pure concept'. They are merely abstractions, useful in helping us to understand the nature of cognitive experience. When we formulate an observation sentence, in that very act we are enlarging upon what is given and relating it to further possibilities – to a whole pattern of past and future, actual and possible experience. Concepts place what is given into a pattern of experiences; the immediately given is meaningless. Regardless of the objections which can be raised about specific features of Lewis' work, the central point of his argument holds: The observation and the theoretical *languages* are part of the conceptual dimension of experience.

Thus, the double-language aspect of the standard account required a new justification, and this was provided by the conceptual–theoretical distinction, as can be seen most clearly in Carnap's formalization. Before turning to his analysis, though, I want to stress that the theory of meaning of the standard account did not change; it remained 'reductionist'. The only required alteration was to include the observation language in the conceptual dimension of experience. However, given the conceptual–theoretical distinction, such an observation language can remain 'theory-independent' or 'theory-neutral'. That is, the same observation language can provide a basis for reduction, and thus for comparison, for many different theories. Having a common, 'neutral', observation basis was the key to meeting the goals of the 'standard account'. From the classical attempts at reduction, the phenomenalist 'sense-data' analysis and the physicalist 'thing-language' analysis, to Carnap's introduction of 'correspondence rules' and 'meaning postulates', the reductionist theory of meaning was designed to meet three important goals: (1) the provision of empirical meaning for non-observational terms, (2) the provision of a means for empirical verification of non-observational claims, and (3) the provision of a standard for demarcation between what is scientific and what is not. A major problem with these attempts at reduction is that so much of the language of science did not turn out to be 'scientific'.

1.2 The 'double-language' view

Carnap's formal analysis had greater influence in modifying the standard account than Lewis' analysis because it raised the discussion of the 'semantic plane', thus avoiding the need for discussion of the specific nature of the given and its relationship to empirical knowledge. For Carnap, a conceptual structure is a language, and the structure of languages, scientific language in particular, can be analyzed formally. He construed the conceptual–given distinction as that between what we want to say about given experiences and the given experiences themselves. We have an infinite number of ways of ordering, understanding, organizing, etc. our given experiences and the differences among these ways can be revealed and examined by analysing the formal structure of the 'linguistic frameworks' in which they are embedded.

Initially, all linguistic frameworks are uninterpreted formal calculi consisting of: (1) predicates for the kinds of entity to be discussed, e.g., 'number' to be used in such sentences as "One is a number," (2) a particular style of bound variable for each kind of entity to range over the entities in question, and (3) a number of analytically true 'meaning postulates' which show the relations of synonymy among the primitive descriptive constants of the framework. If a structure is to have empirical significance, such as the theoretical and the observation languages of science, it must be given an empirical interpretation in a manner consistent with the reductionist theory of meaning.

Stated formally, the theoretical–observational distinction is as follows. The observation language (L_o) is represented by an uninterpreted formal calculus whose empirical meaning is completely determined by its relationship to given experience. The terms which constitute the descriptive vocabulary (V_o) of L_o designate "observable properties of events or things," or "observable relations between them." Thus, interpretation of L_o involves no reference to any theoretical language (L_t). On the other hand, L_t is represented by an uninterpreted calculus which is a "freely

Table 1. Table of dichotomies.

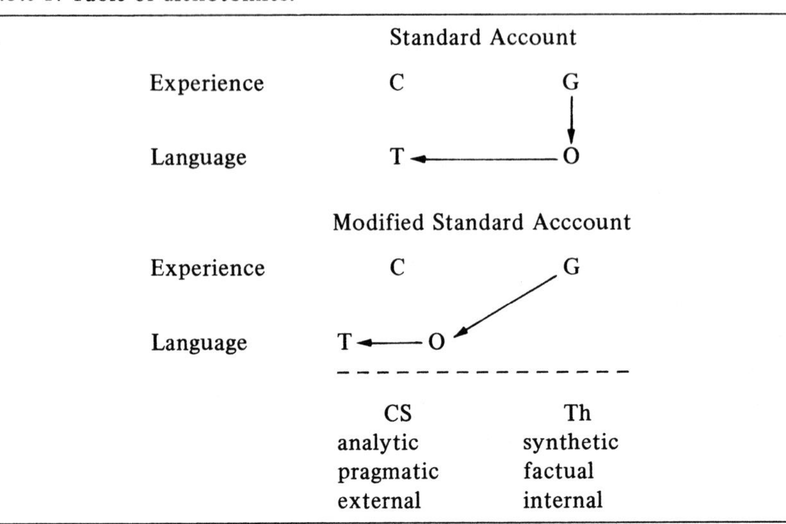

Legend: Arrows indicate direction of flow of empirical meaning. C = Conceptual; G = Given; T = Theoretical; O = Observational; CS = Conceptual Structure; Th = Theory.

floating network." It derives any empirical meaning it has from its connection with some L_o. Correspondence rules (C) link the descriptive terms V_t of L_t with the terms V_o of L_o. Given the rather loose nature of the 'reduction' of L_t to L_o, the interpretation of L_t is indirect and incomplete. What L_o is used is a matter of choice. Different L_o's can be used in conjunction with a particular L_t, such as a 'thing language' or a 'sense-datum language'. More importantly, though, the same L_o can be used as a reduction basis for different L_t's, thus providing a common basis for comparison (see Table 1).

Now, L_o can function in this way only if a conceptual structure does not embody a theoretical viewpoint, i.e., only if we maintain that there is a difference *in kind* between a conceptual structure and a theory. Thus, it is the conceptual–theoretical distinction which now supports the theoretical–observational and, consequently, the double-language view. That is, observation sentences are theory-neutral because L_o does not contain any theoretical hypothesis about the nature of physical reality; the postulates (T) of a theory are contained in L_t. So, while observation sentences are conceptual expressions, and thus involve interpretation, they are not part of any theory. Comparison of different theories is possible by reducing their sentences to a common, neutral observation language.

The essence of the conceptual–theoretical distinction is that a conceptual structure is definitive in nature, while a theory is factual. Thus, the distinction is based upon a traditional analytic–synthetic distinction which Carnap called the 'internal–external' distinction. Internal questions are settled by using the methods appropriate to the framework in question, while external questions are settled by examining the framework as a whole from an outside vantage point. With empirical frameworks, the distinction is that between questions about the adoption or abandonment of parts or the whole of a conceptual structure and questions about the acceptance or rejection of an empirical hypothesis. The latter are settled *internally* by appealing to factual considerations while the former are settled *externally* by appealing to pragmatic considerations. Applicability given certain purposes – not empirical truth – is the key to selection of conceptual principles. Thus, the criterion for the conceptual–theoretical distinction seems to be that experience alone is enough to verify or refute empirical sentences, while conceptual principles are immune to experience and are adopted or abandoned on purely pragmatic grounds.

However, when we ask what are 'purely pragmatic grounds', we see that in a very important way conceptual principles are not immune to experience. If they are to be adopted or abandoned on the grounds of how well they suit our purposes, they are dependent on experience *as a whole.* As Carnap himself phrased it, the question which needs to be asked when considering conceptual principles is: " 'Are our experiences such that the linguistic framework will be expedient and fruitful?' "[5] That is, is the *totality* of our experience such that the principles in question suit our purposes? Thus, the criterion for the distinction between a conceptual structure and a theory is that conceptual principles must answer to the whole of experience while sentences of a theory are accepted or rejected on the basis of individual experiences. But, this is a rather naïve conception of the relationship between a theory and empirical evidence. The arguments against the modified standard account center on this relationship. Those of Duhem and Quine attack mainly the conceptual–theoretical distinction, since its collapse brings about the collapse of the theoretical–observational. Feyerabend, on the other hand, focuses on the theoretical–observational, since his underlying assumption is that every conceptual structure, or 'world-view', embodies a theoretical view-point. These arguments will be discussed in the next chapter. In combination, they have been seen as establishing the 'meaning variance thesis' and thus as presenting us with the 'problem of incommensurability'. I will argue that in reality they only refute the reductionist theory of meaning, which is the heart of the modified standard account. To conclude this chapter I will summarize that account.

The dichotomies required by the standard account, and by the modified standard account, are schematically represented in Table 1. The major change is in the justification for the theoretical–observational distinction. The essential feature of both accounts remains the same, though: Human experience contains within it an independent given element, which provides the foundation for empirical knowledge. The empirically given is shared between theories by its being expressed in the observation language of science, but how it is so expressed is left unanswered. Comparison of different and competing theories is possible because there is at least a 'core' of common meaning embedded in the observation language and that 'core' remains invariant throughout changes in theory. Empirical meaning is supplied to the theoretical language by 'reduction': Theoretical sentences reduce to observation sentences, which in turn reduce to sensory experience. The success of this conception depends, however, on two important considerations: first, that we can, at least in principle, carry out the 'reduction', and, second, given the Lewis–Carnap modification, that we have a viable criterion for the conceptual–theoretical distinction.

CHAPTER 2

'Meaning variance' and 'incommensurability'

2.1 The 'network' view

The network view of meaning has its origins in the objection Duhem raised to the standard account's conception of the relationship between a theory and empirical evidence; in particular, to its failure to take into account the pragmatic element in the construction of theories.[6] In the previous chapter we saw that the key to the conceptual–theoretical distinction is the notion that there is a difference in kind between the acceptance or abandonment of conceptual principles, which is pragmatic in nature, and the verification or refutation of empirical hypotheses, which is factual in nature. Thus, the conceptual–theoretical distinction is based upon what Quine has called the 'pragmatic–factual' distinction. We will turn to Quine's arguments shortly, since for our purposes they are more important given that his focus is on meaning while Duhem's is on verification and refutation. But first, I will give a concise formulation of Duhem's argument because it did strongly influence Quine in his formulation of the network view.

Duhem argued that the pragmatic–factual distinction made by the standard account is at best simplistic. As noted previously, if conceptual principles are to be adopted or abandoned on the grounds of how well they suit our purposes, they are dependent upon experience as a whole. In supposed contrast to this, the sentences of a theory are to be accepted or rejected on the basis of individual experiences. Consider, however, the case of empirical laws. Once an empirical law is well established the tendency is to ignore or try to accommodate recalcitrant experiences, rather than give up the law. The history of science is replete with examples where apparently falsifying evidence was ignored, swept under the rug, or led to something other than the law being changed. It is more in accordance with scientific practice to say that experience is not sufficient to refute a single, individual theoretical sentence. In the face of recalcitrant experience, something must go, but there is some choice involved. There is no clearly defined dividing line between empirical refutation and pragmatic abandonment. What is involved in testing an empirical hypothesis is that *the whole body* of sentences that make up the theory, some of which may be considered conceptual principles, others empirical hypotheses (Quine even adds the principles of logic and mathematics), faces the 'tribunal' of experience. In making changes, we do seem to adhere to a 'conservation principle', though, in that we first try to make changes where the effect will be felt the least. As Quine has expressed it, using Neurath's apt metaphor: 'Fixing' the network is like fixing a ship at sea – we make changes in such a way that we do not sink the ship.

The most important consequence of this conception of the relationship between empirical evidence and theory is that if meaning and verification are co-extensive, then, if evidence is applicable only to the whole theory, so is meaning: This is the central thesis of the network view of meaning. As Quine expressed it:

> The meaning of a sentence lies in the observations that would support or refute it. To learn a language is to learn the meaning of its sentences and hence to learn what to count as evidence for or against them. The evidence relation and the semantical relation of observation to theory are co-extensive.[7]

So:

> It is no shock to the preconceptions of Old Vienna to say that epistemology now becomes semantics. For epistemology remains centered as always on evidence, and meaning remains centered as always on verification. What is likelier to shock preconceptions is that meaning, once we get beyond observation sentences, ceases in general to have any clear applicability to single sentences.[8]

Although Quine is in fundamental disagreement with the sharp dichotomies of the standard account, its belief in the co-extensiveness of meaning and verification, and thus its reductionist theory of meaning, forms the cornerstone of his version of the network view. Let me back up some now and address the issues under discussion from Quine's perspective.

Quine maintains that the general belief that there is a difference in kind between 'questions of language' and 'questions of fact' is at the heart of the conceptual–theoretical distinction. Since meaning and verification are co-extensive, there are two criteria, which Quine calls the "two dogmas of empiricism," for making the distinction: the analytic–synthetic distinction and the pragmatic–factual distinction (which is the essence of the "dogma of reductionism"). These criteria function to distinguish a conceptual structure from a theory in the following way. For a *conceptual structure*, the sentences, which show the relations of synonymy among the primitive descriptive constants (Carnap's meaning postulates) and which, thus, articulate the criteria for the application of terms, are *analytic*. The question of adoption or abandonment of a structure is settled *pragmatically*, from an *external* vantage point. On the other hand, the sentences of a *theory* are *synthetic*, i.e., have factual content. Questions of verification or refutation are settled on *factual* grounds alone, using the methods appropriate to the framework, i.e., they are settled *internally*.

Quine argues that the analytic–synthetic distinction is a dogma, inconsistent with the principles of empiricism. His main argument is that analyticity is an intensional notion and can be defined only in terms of other intensional notions. However, intensional notions cannot be defined extensionally, i.e., in terms of empirical criteria. The analytic–synthetic distinction rests upon the notion of "sameness of meaning," which cannot be defined empirically. If we examine the process of translation in a truly empirical fashion, i.e., by means of behaviorist field linguistics, and study situations of "radical translation," we see that we can make no empirical sense, and thus no sense, of the notion beyond "similarity of stimulation." Thus, the distinction has no empirical foundation. Quine's argument is, of course, squarely based on the acceptance of behaviorist linguistic theory, which has been rejected by most linguists. There are accepted ways, among linguists, of making an analytic–synthetic distinction and of distinguishing questions of language from those of fact, but it is not necessary for us to become embroiled in these issues. It is enough to see that his adoption of behaviorist linguistics is just another expression of his acceptance of the reductionist theory of meaning, with the 'given' now being 'sensory stimulations'. This will become more evident when we consider Quine's attempt to salvage his 'ship' in the next section.

Quine's objections to the pragmatic–factual distinction are more central to the argument against the standard account and in favor of the network view. His points are basically those of Duhem: (1) In an important sense conceptual principles are not immune to experience and (2) There is a strong pragmatic element in the acceptance or rejection of the components of a theory. So, even if it were possible to make some distinction between a conceptual structure (linguistic framework) and a theory, it could not be the one provided by the standard account.

Quine's own contribution comes in seeing the distinction as related to the "dogma of reductionism." In Section 1.1, we saw that the standard account has two main aspects: a double-language conception of scientific theories and a theory of meaning which requires reduction of those languages to sensory experience. The specific feature of the reductionist theory of meaning, which Quine calls a "dogma," is the notion that *individual* sentences are related to sensory experiences. Evidence applies to a theory as a whole, so a one-to-one reduction is not possible, i.e., no particular sentence can be reduced to *its* empirical content. Rather, the sentences of a theory form an interwoven network: They range from the highly theoretical to the highly observational, with no sharp dividing line. Thus, the sharp theoretical–observational distinction behind the double-language conception is replaced with what Quine calls a "graded" distinction. However, the theory of meaning behind the network view is still 'reductionist' in that any empirical meaning a network is to have must come from the sensory stimulations which 'impinge' upon its boundary. This meaning is 'reflected' in the observation sentences and is distributed throughout the network by whatever meaning connections there are among all of its sentences. Just how different Quine's notion of an observation sentence is from that of the standard account will be discussed in the following section.

Before moving on to a discussion of the problems inherent in the network view, let me summarize it by looking at the feature that is central to those problems: the relationship between the whole of a network and its parts. For the reasons just discussed, the conceptual–theoretical distinction collapses and the whole network is considered to be a theory in much the same sense as any 'scientific' portions of it. In both cases the issue is what is the best way of understanding, predicting, controlling, etc., our sensory stimulations. The whole network forms the background within which we conduct scientific research. Because it is a 'net', each part is connected with all of the others. Change in a part, e.g., in physical theory, causes reverberations throughout the whole. Most importantly, only in relation to the network is it possible to say what meaning connections there are. Additionally, the relationship between a part of the network and the whole makes the Carnapian internal–external distinction untenable. Changes cannot be made from an 'outside' vantage point. We must work within the network to bring about change. Quine maintains that it is part of our biological make-up not to make radical changes unless necessary: We warp meaning gradually in order not to 'sink the ship'. However, when we try to make sense of the holistic conception of meaning of the network view, we run into serious difficulties. What happens when we do make radical changes, such as the change in physical theory from classical mechanics to relativistic? Do we then have *two* 'ships'? 'Marooned at sea'?

2.2 Shipwrecked

When the implications of the network view of meaning and verification are followed out, we are led unrelentingly to the conclusion that different networks are incommensurable. That is, if we take the whole–part relationship seriously, then a difference in physical theory, for example, creates different networks because of the difference in meaning connections throughout the whole of the network. This result has been called the 'meaning variance thesis'. The significance of the thesis is that the sentences of a theory become meaningless intertheoretically. So, different physical theories cannot contradict or agree with one another. They cannot be consistent or inconsistent with one another. There is no neutral observation language in which to couch the results of a 'crucial experiment', so as to provide a common basis for comparison. The situation becomes even worse when we recall that the network view is part of the more comprehensive Frege–Russell theory of meaning in which sense determines reference. Thus, it would seem that different theories are about different 'realities': The 'world' of Newton is not the 'world' of Einstein. And – to add the final straw – there can be no mutual standards for choice between competing theories, since the standards which must be met by a theory are relative to its network, i.e., are 'internal'. There are no objective 'external' standards for choice between two equally comprehensive theories, each true to the standards of its own network.

All of these consequences of the network view have been lumped together under the title of the 'incommensurability problem'. Quine, however, did not raise the problem himself. It was revealed by Kuhn and Feyerabend independently in the same year.[9] Quine's own attempts to avoid the 'problem', which he maintains is an expression of an unacceptable "epistemological nihilism," center on resurrecting the notion of the given in an acceptable form.[10] He has tried to construct a "graded" notion of observation sentence, where observation sentences at the lowest level are direct expressions of sensory experience and thus escape the network. Such sentences are clearly inconsistent with the network view. In his own argument against the "dogma of reductionism," Quine maintained that the only distinguishing feature of observation sentences is a particular "germaneness" to sensory experience. Since change causes reverberations throughout the network, we try, first, to make changes where they will be felt the least – in the observation sentences. However, if these sentences must perform the task assigned to them by the standard account, i.e., if they are to function as intertheoretical "arbiters," there has to be a way of detaching them from the network of which the physical theory is also a part.

In order to detach the lowest level observation sentences so that they can perform the function of arbiters, Quine attempts to develop a behaviorist account of perception. This account depends, however, on maintaining a 'third dogma': stimulus–response psychology. Here the observer is considered to be like a photoelectric cell or a measuring instrument, uttering a string of sounds in response to sensory stimulations. Given that the sensing apparatus of human beings is fairly standard, we can assume that our stimulations are similar enough. So, the lowest level observation sentences are directly related to something outside the network that we all share. Quine is satisfied with this account because, as thorough-going empiricists, we should fuse empirical psychology with epistemology. However, on the surface it seems rather strange that the psychological theory he adopts is the most crude version of behaviorist psychology and that he adopts it without even considering the objections raised against it by empirical psychologists themselves. The puzzle disappears when we make the connection between Quine's version of the network view and the standard account: the reductionist theory of meaning. Any empirical meaning a theory is to have must come from the observational situation. Thus, we must come to the observational situation in as neutral a state as possible in order to make as direct contact as possible with an independent physical reality. At the lowest level, our sensory stimulations constitute the meaning of an observation sentence: ". . . it has an empirical content all its own and wears it on its sleeve."[11] Quine has simply replaced the 'given' of the standard account with the 'triggering of nerve endings'.

The scientifically acceptable version of the reductionist theory of meaning, according to Quine, is as follows: (1) What is given and shared are sensory stimulations, i.e., triggerings of nerve endings; (2) Observation sentences at their most basic level obtain meaning directly from sensory stimulations; and (3) Once we get beyond this level of observation sentence, meaning has no clear application to individual sentences, but only to groups of sentences or to a theory as a whole.

Now, there is a sense in which Quine is right. If we are to interpret the reductionist theory of meaning in a way that is acceptable to modern empiricists, a more exact, scientific conception of the given must be formulated. The problem is that even if it were possible to formulate such a conception, these lowest level observation sentences will not solve the 'problem'. "Similarity of reception" and "similarity of perception" are not the requisite notions for intertheoretical comparisons. In Quine's own terms, "community of agreement" – what elsewhere I have called "similarity of assent" – on observation sentences is needed for commensurability.[12] But, "similarity of assent" requires relativizing the notion of an observation sentence to a language community, i.e., we need to have a similarity of language in which to express our observational claims. Given the holism of the network view, those who hold different theories employ different networks, i.e., are not part of the same language community. This, in turn, concedes the point to the "nihilists": "Similarity of assent" is not a meaningful *inter*network notion.

So, the situation seems as follows. If we agree that the sharp distinctions made by the standard account are untenable, we are stuck with the 'incommensurability problem'. Indeed, a great deal of philosophical energy over the last twenty years has gone into trying to find a way around the 'problem'. Before briefly discussing the most influential attempts, I want to restate the problem situation for those philosophers in Feyerabend's terms, since it is his formulation to which most were responding. My discussion will focus on the early Feyerabend and primarily on one critical point made by him. His position has become more incoherent over time and this has had the unfortunate consequence of obscuring his important early argument. Its force has not been sufficiently appreciated – not even by him! His main argument is that the reductionist theory of meaning of "orthodox empiricism" (standard account and Quinian) is untenable. From this he concludes that incommensurability is a *fait accompli*: All that is possible is an "anarchistic" theory of knowledge. But, his conclusion follows only if the reductionist theory of meaning is the only theory possible. All that has actually been shown is that the reductionist theory is a *theory* which has been adequately refuted.

The central objection Feyerabend raises against "orthodox empiricism" is that at its foundation there is the *a priori* assumption of a special relationship between the human observer and an objective, i.e., observer-independent, physical reality. This *a priori* assumption is the heart of the theoretical–observational distinction. Feyerabend labels any position which makes this assumption a "semantic theory of observation."[13] A "semantic theory" holds that the sensations or sensory stimulations of the human observer have a significance for meaning. Observation sentences are distinguished by receiving their empirical meaning directly from the observational situation, independent of any theorizing. The semantic theory is, then, the core of the reductionist theory of meaning. However, Feyerabend contends, a truly empirical philosophical position must see the interaction between the observer and the 'world' as one of the processes which needs to be studied: We cannot legislate its nature *a priori.* The assumption made by orthodox empiricism should really be considered an empirical hypothesis and thus be subjected to critical examination; in particular, we need to examine it in the light of other hypotheses. Thus Feyerabend proposes one, which he claims is more in accord with empirical research concerning human observers and their relation to the 'world'. He adopts the perspective of Gestalt psychology: Human observers are not passive spectators in the arena of experience; rather, experience, in any meaningful sense, must occur within a comprehensive 'world-view' (conceptual structure) and that world-view is part of the disposition we *bring* to experience. Since any sufficiently comprehensive world-view is a theory, all observation sentences are theory-laden.

Feyerabend does not deny that there are distinguishing criteria for observation sentences. On the contrary, he argues that a definite causal relationship can be seen to exist between sensory experience and certain utterances. He offers an alternative characterization, which he calls a "pragmatic theory of observation": Observation sentences are distinguished simply by the context in which they are most likely to be uttered.[14] They do not, however, get their meaning from the observational situation. The observer, in uttering such sentences, should be considered to be a measuring instrument, making a string of sounds in response to stimulation. This view is similar to Quine's, but the similarity stops here. Feyerabend contends that it is a mistake to turn this "pragmatic theory" into a thesis about meaning. Taking the measuring instrument analogy in its fullest sense, he maintains that just as pointings of a needle on a dial are in need of interpretation, so too are human observational utterances. Observational sentences can only be interpreted by theory, so theoretical sentences cannot be reduced to them for their meaning. Although there is some ambiguity in what he means by 'interpretation', we can see his point. It is theory which gives meaning to observation sentences and not the other way around. This is Feyerabend's version of the network view. He calls it "theoretical realism."

Theoretical realism has three basic components: (1) A pragmatic theory of observation, i.e., observation sentences are distinguished only by the context in which they are most likely to be uttered; (2) the meaning of an observation sentence is determined in full by the theoretical context in which it is embedded; and (3) theories have meaning independent of observation. It maintains a *thoroughly* holistic view of meaning. All the sentences of a world-view are theory-laden, with all the principles of a theory contributing to the meanings of the terms within it. Taken in isolation sentences mean nothing. Thus, the 'meaning variance thesis' follows. Since all the principles of a theory contribute to meaning, change in principles entails change in meaning. A slight change might not be pernicious, but any significant change would be. The fact that words in sentences of different theories are typographically the same and that the sentences seem to follow the same grammatical rules is not helpful: Meaning and reference change with changes in network connections.

We can see immediately that when attempting intertheoretical comparisons, the 'incommensurability problem' arises. There is no meaning contact between different theories, since the 'meaning connections' for what appears to be 'the same' word, perhaps even in 'the same sentence', are different. We cannot compare theories by reference to 'experience' either, even when they seem to concern the same subject matter, for sense determines reference. So, the things referred to by two different theories are themselves different. That there is a 'same subject matter' requires that there be theory-independent experiences to capture in a neutral observation language, providing 'the facts' for all theories in that domain. However, if we observe and experience only from the standpoint of a theoretical network, the subject matter of different theories is also different. The claim that different theories make different claims about the same entities, thus providing a basis for comparisons, assumes that we have some extra-network way of identifying the entities, which, as we have seen, is inconsistent with the network view. Incommensurability results because there is no way of distinguishing, epistemologically, between the 'ship' and the 'sea'. Any network-independent 'given' that could be located would be too 'pure' to be of any use. So, 'reduction' is not possible at all, since what would there be to 'reduce' to? The reductionist theory of meaning has been 'sunk'.

2.3 Is there meaning after Feyerabend?

The obvious conclusion by now should be that the reductionist theory of meaning has failed, so we need to construct a new theory of meaning for scientific theories. Feyerabend's own conclusion is that incommensurability cannot be avoided, and, moreover, that this is substantiated by an examination of the history of science.[15] One major problem with Feyerabend's conclusion, though, is that he never makes clear what he means by 'meaning'.[16] But, it is necessary to do so before we can draw such conclusions from the history of science as, e.g., 'mass' in classical mechanics and 'mass' in relativistic mechanics do not 'mean' the same thing. The only thing he says about meaning is that it "comes from" the theory in question. He has never attempted to develop his own theory of meaning; indeed, from the way he has responded to questions on this issue it seems that he at least tacitly assumes that the reductionist theory is the only possible theory, and since it has failed, we have no alternative but to live with incommensurability.

Although there has been a widespread acceptance that the reductionist theory is not viable, while there is much to be said in favor of a network conception of meaning, few philosophers have followed Feyerabend's path – though many have worked on the 'problem' he presented. In general, two approaches have been taken: (1) to attempt to demonstrate that, in contrast to Feyerabend's assessment of historical data, detailed studies of actual theory change reveal that incommensurability does not pose a serious problem and (2) to attempt to find a way around the incommensurability of meaning of the network view through an examination of the nature and necessities of language.[17]

Let me begin with the second approach. One possible way around the 'problem' lies in the realization that it arises not only because of the failure of the reductionist theory, but also because there is a more general theory of meaning in the background of that theory: the Frege–Russell theory in which meaning has two components, 'sense' and 'reference', or, in Carnap's terminology, 'intention' and 'extension', and it is sense which determines reference. If there could be a way to determine reference independently of sense, so that at least a part of the extensions of the theories we wish to compare would overlap, the 'problem' would at least be lessened. This approach has been tried by several people, beginning with Scheffler.[18] However, these attempts miss the point: As long as we accept the Frege–Russell theory, the task is hopeless. The only way we have of identifying the reference of a word is by the criteria formulated in its sense; so, even if it were true that the reference remained the same when the sense changed, we would have no means of *knowing* it. We cannot identify an entity independent of our description of it.

Kripke and Putnam, independently and for different reasons, presented a similar alternative to the Frege–Russell theory: the 'causal theory of reference' (Kripke) of the 'causal theory of meaning' (Putnam).[19] Although there are differences between the two versions, for our purposes we can ignore these and focus on the similarities. Kripke's initial insight came in response to problems having nothing to do with meaning in scientific theories: problems with the Frege–Russell theory of descriptions and proper names. He has since extended it in such a way that it is applicable to the 'natural-kind' terms and theoretical terms in scientific theories.[20] The central notion of the 'causal theory' is that of a 'rigid designator'. Certain words, such as proper names, function in such a way that, although their referents may not satisfy any of the descriptions thought to be true of them, they still pick out that referent, i.e., they designate it 'rigidly'. For example, John Jones may not satisfy any of the descriptions I believe of him and yet he is still the referent of 'John Jones'. The referents of such words are fixed initially by ostension and are continued in time by a 'causal chain', i.e., an 'historical chain', linking the users of the term. This works quite well as a theory of proper names, but whether or not it works with scientific terms is another issue.

Putnam's version of the causal theory grew out of a concern with incommensurability and scientific theories. His discussion of natural-kind terms is similar to that of Kripke regarding proper names. The part of his version which concerns theoretical terms is 'causal' in two respects: First, there is a causal relationship which connects the referent of a theoretical term with an observable phenomenon, i.e., the referent of a theoretical term is the *cause* of an observable effect; and second, there is a causal (historical) relationship between referers using the term. The causal (historical) aspect of the Kripke–Putnam theory works in the following way for natural-kind terms and theoretical terms. It is claimed that there is an introducing or 'baptismal' event in which the referent is fixed either by ostension, usually supplemented with a description, or by giving a causal description of the referent by picking out the effects for which it is thought to be responsible. The descriptions given at this 'event' need not be true. True descriptions can only be found by scientific investigation into the 'essence' of the referent. Communication takes place between the users of the term because of the 'stereotype' associated with each word, so it is not necessary to know a description, and because we understand the relation 'same kind'.

We can easily see the attraction of the causal theory for philosophers of science since, unlike traditional theories of meaning and reference, it was designed to account for diachronic as well as synchronic aspects of language.[21] Although it assumes, partially in accordance with the Frege–Russell theory, that the presence of 'essential properties' can function as a criterion for the application of a word at a particular time (synchronic aspect), the causal element of the theory provides for theory-independent reference over extended periods of time (diachronic aspect). The causal aspect can be illustrated as follows. At some point in time the 'stuff' in lakes, rivers, in which we bathe, which we drink, etc. was baptized 'water', and from that moment on that stuff is rigidly designated 'water' no matter how our beliefs about it (or about its essence) may change. (In current terminology, the stuff in question is "rigidly designated 'water' in any possible world.") Thus, the incommensurability problem is solved – of course only in respect to reference – because, despite how we may change our views about the nature of *water*, the word 'water' will always refer to that *stuff*, and therefore we can compare different theories about it. So, the causal, diachronic aspect of the Kripke–Putnam theory commits us to the name of a natural kind, but not to a conception of its essence. One major problem with this feature of the theory is that reference is too rigid: What do we do with terms, such as 'aether', which were once thought to refer, but are later thought not to?[22]

The synchronic aspect of the causal theory contains its most problematic feature: the thesis of essentialism. Kripke and Putnam maintain that the goal of science is the discovery of the essence of natural kinds, since "things are what they really are." Once we find out, for example, that the essence of water is H_2O, we will henceforth refuse to call anything 'water' which is not H_2O. Using the famous Putnam example, if we were to find out that the stuff we have been calling 'water' on Twin Earth is really XYZ and not H_2O, then we would withdraw the name, since "nothing counts as a possible world in which water is not H_2O."[23] So, once we have found the essence, our commitment to the name is supplemented by a commitment to a conception of the essence. The obvious problem with this argument is how do we know when we have found the essence? Is it possible to formulate a criterion for deciding whether a given scientific theory about the nature of a natural kind reveals the *real* essence?[24]

A more fundamental problem has recently – and quite forcefully – been pointed out by Shapere: The facts of scientific practice are not consistent with the essentialism and the rigidity of reference of the causal theory.[25] The problem is "more fundamental" in that the force behind the argument stems from a disagreement in *method*, i.e., a disagreement about how one can arrive at a theory of meaning for scientific theories. So, even though there may be other versions of the causal theory, some more adequate than others and some perhaps not subject to all the criticisms Shapere makes against the Kripke–Putnam version, it is not necessary to consider them since the crucial objection is to the *approach* taken by these theories. All versions assume that from a study of the necessities of language we will find out what scientific language must be like. This attitude is underscored by the frequent use of 'science fiction' examples to illustrate points being made: The examples are just that – illustrative – and are as unnecessary to the formation of a theory of meaning as is the examination of 'scientific fact'. Whereas Shapere argues that only a study of scientific practice and the history of that practice can lead to an understanding of the nature of science in general and, in this specific case, of its language. A kindred view forms the premise upon which Parts II and III of this book are based: Although this is a book about meaning, it is not about language *per se*, but about science and features of its language. It is about the practices which are part of the creation of meaning in scientific theories.

Now, discussions of scientific practice have been used repeatedly over the last twenty-five years; e.g., in the arguments which brought about the collapse of the distinctions of the standard account and which led to the network view; in Feyerabend's arguments for incommensurability and meaning variance; and in various 'case study' responses to incommensurability. However, the study of scientific practice has not been put to its fullest use. Examples of scientific practice have been considered of value – which in itself is an improvement over the standard account – but mostly in a *negative* way, i.e., as counterexamples to show how certain philosophical conceptions are not true of science or not useful for understanding science. The study of scientific practice has not been exploited fully in the *positive* sense, i.e., it has not been seen as forming a significant portion of the source of our conceptions about various aspects of science; in the case at hand, as the source of a new theory of meaning.[26]

It is rather strange that we now let scientific practice count *against* our theories of meaning, but not *towards* the construction of a theory. We act as though we believe such a theory could be arrived at *a priori*, or at least in a way independent from an examination of its subject. But this is precisely what has happened. The 'linguistic turn' in philosophy has led philosophy of science away from an examination of its subject – science. The survey of the history of philosophical conceptions which has been provided in these two chapters has shown that one of the central problems in the philosophy of science for nearly three-quarters of a century has been the nature of meaning in scientific theories. Yet, there is still no adequate conception of meaning in scientific theories. Why? The answer to this question lies in seeing the "mistake" common to the theories which have been proposed thus far: ". . . the supposition that the nature of science can be illuminated by an examination of alleged necessities of language which are independent of the results and methods of scientific inquiry."[27] This is why, in the causal theory, science fiction examples are deemed as good as examples drawn from real science. This is why, in the standard account, the 'context of discovery' (meaning in this case the examination of how concepts actually arise and develop in scientific practice and how, in general, scientific language develops) is irrelevant. The ahistorical nature of the standard account and the science fiction examples of the causal theory are but two sides of the same coin. Both positions have been formulated on the assumption that study of how meanings are formed, developed, and used in scientific practice is irrelevant to the philosophical enterprise.

That the 'linguistic turn' was the wrong turn for the philosophy of science is a criticism that was made by Popper near the beginning of our story. However, in response, he banished the study of meaning from the philosophy of science. This is why he has not appeared in this 'mini-history', although my emphasis on problems and problem situations has quite obviously come from the influence of his thinking on my own. Popper sees the standard account's emphasis on meaning as an attempt to avoid talking about 'truth' because its proponents recognized that some very good science, such as Newtonian mechanics, cannot be called 'true'. So, the criterion of 'demarcation' between science and metaphysics could not be given in terms of 'truth', but rather, in terms of 'meaningfulness' and 'meaninglessness'. In response to this Popper argues that it is not 'being reducible to sensory experience', but 'having testable consequences' that makes a theory 'scientific'. Thus, his focus has been on his criterion of 'falsifiability' and its consequences, to the exclusion of questions concerning meaning.

Whatever problems there are with the tradition we have been examining and whatever the initial motivations behind the path taken, there is a "truth" behind its efforts: The creation of meaning is a major aspect of the scientific enterprise. So much of that enterprise concerns the introduction and refinement of ways of conceptualizing 'the world'. Formulating an understanding of this aspect of science is an important problem and one which should be undertaken by philosophers. The trouble with the views we have been considering is not with their problem, but with their *approach*. The real "truth" of the matter is that the nature of meaning in scientific theories must be seen in the context of *the network of beliefs* (theoretical, methodological, metaphysical, common sense) and *problems* (theoretical, experimental, metaphysical) which is part of the making of meaning in scientific practice – of the introduction and development of the concepts and the terminology of theories. Only through study of the practice of meaning construction can we come to an adequate understanding of the very notions of 'meaning' and 'meaning change' as they relate to science.

Additionally, although the study of scientific practice is necessary to the formulation of this understanding, it is not sufficient. We should also make use of valuable insights from those sciences which study meaning and concept formation: linguistics and cognitive psychology. These sciences can assist us in clarifying issues, such as what it means to 'have' a concept, and can also assist us in articulating a framework in which to understand meaning and concept formation in science. Notions developed in these fields cannot, of course, simply be imposed upon, or applied in a wholesale manner to, science; rather, those notions which seem useful for understanding the scientific situation should be adopted and adapted to fit it. Thus, in contrast to the approach taken by the views surveyed here – views which have dominated our thinking about meaning in scientific theories for so long – the approach I am advocating is one which is firmly rooted in extensive study of scientific practice and coupled with insights from the scientific study of meaning in 'ordinary' as well as scientific contexts.

PART II

The Scientific Situation: An Historical Analysis

What has perhaps been overlooked is the irrational, the droll, even the insane, which nature, inexhaustibly operative, plants in an individual, seemingly for her own amusement. But these things are singled out only in the crucible of one's own mind.

Einstein

Introduction

In this Part I will present an analysis of the formation of a scientific concept: the concept of 'field', beginning with Faraday's attempt to formulate a continuous-action conception of electric and magnetic actions and ending with Einstein's formulation of the special theory of relativity. Thus, our concern is with the electromagnetic field concept only and not with the gravitational and quantum field concepts. Our purpose allows us to stop with the special theory of relativity, since it is there that the modern concept of field, as an 'independent reality', appears. What this means will emerge from the discussion. Also, an analysis is provided of only the contributions of Faraday, Maxwell, Lorentz, and Einstein. To give a 'finer' analysis would require several volumes. Fortunately, this is not necessary for our enterprise since, although there were many contributors to the formation of the field conception of electric and magnetic forces, the most significant changes in *meaning* were made by these four.

A few preliminary points should be made about the approach taken. First, the analysis is based primarily on source material. There exists a vast literature on various aspects of the development of electromagnetic theory, some of which is excellent and some of which is relevant to my analysis. Wherever discussion of that literature seems necessary, it is relegated to the footnotes. Second, even the analysis of the four major contributors is incomplete and some interesting questions related to this discussion are simply noted and postponed for future works. Hopefully, sufficient details have been discussed to establish the fruitfulness of this approach for understanding the nature of meaning in scientific theories. Third, although the analysis of each phase of development is necessarily brief, I have tried: (1) to provide some of the network of methodological, theoretical, and metaphysical beliefs of each scientist considered, (2) to indicate the problems that figured significantly in the attempts to articulate a field conception of electric and magnetic forces, and (3) to discuss what reasons were given for, or lay behind, each significant change in meaning. Finally, I have attempted to make the analysis as non-technical as possible in order to make both the historical material and the discussion of it accessible to a wider audience.

The concept of electromagnetic field has been chosen for several reasons. I wanted to begin with a period in which there was already an established scientific community and to discuss the introduction of a new scientific concept, from its inception to its form in modern physics. The field conception of forces is a good choice since it is one of the – if not *the* – major conceptual innovations of 19th and 20th century physics. Although embryonic field concepts were around before Faraday, the analysis begins with his conception because it was with him that the idea that processes in the region surrounding bodies and charges are *essential* to the description of electric and magnetic actions first appeared and took hold; it is in connection with his work that the term 'field' first came into use. Stated most simply, the concept of field involves the notion that some physical processes take place in the region surrounding the bodies in question; in the case at hand, that electric and magnetic states characterize points in the space *around* charged bodies and magnets. I will call any conception of electric and magnetic actions a 'field concept' if it contains this notion. The central question that arises in connection with a field conception of forces is how, or by what processes, can the forces in question be transmitted continuously through space where there is no 'ordinary' matter. In the period under discussion that question was expressed as: Do these processes take place in space free from all matter, ponderable or otherwise, or are they states of some all-pervasive medium, 'aether', filling space? If the latter, what is the nature of that aether?

A further reason for choosing the field concept is that it spans several 'paradigms': classical mechanics, the theory of electrons, relativity theory, and, if we were to continue, quantum mechanics – and, potentially this could yield some interesting data, relevant to contemporary notions of 'paradigm', 'research program', etc. Finally, as a case of concept formation, it is particularly interesting since what analyses there have been of the development of new scientific concepts have tended to focus on those which have been formed by 'differentiation' from or 'combination' of existing concepts. Although the case does involve some differentiation (from the 'aether', which was also under development in that period), the formation of 'field' primarily involves conceptual 'shifts' with respect to its ontological status: Is it a 'property' (of what), certain 'processes' (in what), or a 'substance'?

It might be helpful to conclude by summarizing beforehand the development of the concept of field as set forth in the following chapters. First, with Faraday's 'lines of force', the field is a physical state of stress and strain in space. The question as to whether or not space is filled with an aether is left open, with Faraday leaning towards the view that the lines of force are themselves substances existing in "mere space." Maxwell, however, interpreted the stresses and strains to be states of a material medium. He conceived of the field as a state of a mechanical aether, i.e., one which obeys Newton's laws. In his fully-developed theory, though, he acknowledged that it might not be possible to formulate an adequate mechanical model of the aether, while at the same time he held that the concept of field is essential to the description of electric and magnetic actions. Lorentz, while maintaining that the field is a state of an aether, altered the nature of both. Lorentz' aether is a non-Newtonian substance. Finally, in asserting that the aether is "superfluous," and going on to establish why, Einstein put the field ontologically on a par with matter. That is, our present concept of field is that it is an *irreducible* element of description. These conceptions will now be discussed in more detail.

Michael Faraday (1791–1867)

CHAPTER 3

Faraday's 'lines of force'

3.1 Initial conception

Faraday was perhaps the ideal candidate for introducing a new world-view. He was a great experimentalist, but he was also strongly inclined towards what he called "speculation." His experiments were crucial in his attempt to understand the nature of electric and magnetic actions, but his dissatisfaction with the action-at-a-distance conception of these actions (Ampère *et al.*) did not stem simply from his experimental observations. Faraday engaged in a great deal of metaphysical speculation, both as a result of his experiments and as a guide to experimentation. Experimentation and speculation, *together*, constituted his research.[1] It must be stressed, though, that he took great care to separate what his experimental data could support from what was speculation, since he was somewhat wary of speculation.[2] While acknowledging its value, he maintained a cautious attitude towards it, which made him hesitant to express his speculations publicly. However, his hesitations diminished some as he gained more confidence in them and developed them more fully. Thus, it is difficult to say *what* his new conception was, in its entirety, and *when* he first had it.[3] The position I will hold is that at least from the time of his "Historical sketch . . . ," which concluded with his discovery of electromagnetic "rotations," i.e., *before* his discovery of electromagnetic induction in 1831, he believed that a new conception of electric and magnetic actions was possible.[4] He tried to work out such a conception during his subsequent researches.

Faraday's attitude towards speculation extended beyond his own. In a letter of 1821 to De la Rive, he said:

> You reproach us here with not sufficiently esteeming Ampère's experiments on electro-magnetism. Allow me to extenuate your opinion a little on this point. With regard to the experiments, I hope and trust that due weight is allowed to them; but these you know are few, and theory makes up the great part of what M. Ampère has published, and theory in a great many points unsupported by experiments when they ought to have been adduced. At the same time, M. Ampère's experiments are excellent, and his theory ingenious; and, for myself, I had thought very little about it before your letter came, simply because being naturally skeptical on philosophical theories I thought there was a great want of experimental evidence.[5]

So we can see that from at least 1821, Faraday held that the prevailing action-at-a-distance conception is a *speculation* and *possibly incorrect.* This attitude set him apart from others working in the area at that time.

He continued the letter with a discussion of his discovery and interpretation of the rotations, to which I will return shortly. But first, we need to discuss how Faraday arrived at his skeptical position and what form his own initial speculation took. Both of these are hotly disputed issues in the Faraday literature. Some writers stress religion, some stress the influence of philosophers, and some try to deny altogether that he had a new conception at this point.[6] What can be safely said is that Faraday reflected critically on the nature of matter and force at least from the time of his early chemical researches. His lectures during the period 1816–1819 are concerned with such topics as: "general properties of matter," "attractive powers of matter," "chemical affinity," etc.[7]

These reflections were provoked by various influences. The primary influence on his opinions remained, throughout his life, his experimental observations; but these alone were not sufficient to produce a new conception. He was a deeply religious man and wished to formulate a unified view of nature, which would be consistent with his religious beliefs. Undoubtedly he was also strongly influenced by the knowledge and views of Davy, his benefactor and mentor. Davy was concerned primarily with the nature of chemical attraction, repulsion, and bonding. He was familiar with the views of Boscovich concerning the nature of matter and seems to have held a position similar to his: that matter is composed of point atoms surrounded by attractive and repulsive forces.[8]

Additionally, as has been pointed out by Levere, there was a British tradition of 'force atoms' with which Davy must have been familiar.[9] Faraday, himself, probably never read Boscovich and did not make reference to him publicly until his 1844 "Speculation . . . ."[10] There he presented his own conception of atoms as point centers of converging lines of force for the first time.[11] At what point he had formulated this conception was not noted. He simply stated that it was the view he held at the time he presented it.

His convictions regarding the unity of forces were expressed at a much earlier point – in his first chemistry lecture of 1816: "That the attraction of aggregation and chemical affinity is actually the same as the attraction of gravitation and electrical attraction I will not positively affirm, but I believe they are . . . ."[12] This view is consistent with Davy's position at the time and with Faraday's religious beliefs in the unity of nature. What it means for the forces to be "the same" was not discussed. Faraday's notion of the 'unity of forces' can only be understood fully in terms of his later-expressed conception of atoms: If atoms are convergent lines of force, if all actions take place through lines of force, and if the lines of force are the only substances, it follows quite consistently that all forces would be unified and interconvertible.[13] Faraday, however, did not even hint at such a conception until a much later time and we would be going too far to attribute it to him at or near the outset.[14]

In concluding this inevitably somewhat vague discussion of the nature of Faraday's early reflections, I want to emphasize this last point again. Faraday did not have a fully articulated field conception of electric and magnetic actions when he began. The 'lines of force' were introduced into the description of electric and magnetic actions after his discovery of electromagnetic induction – as a second attempt at explanation of the phenomenon – and only much later was it hinted that they might afford a unified description of all forces. His field conception, as we will see, was initially quite vague, was formulated over time, and was never fully articulated.

3.2 Electromagnetic rotations

The first important experimental influence on Faraday's thought was his discovery of electromagnetic "rotations."[15] In the process of writing his "Historical sketch . . . ," Faraday repeated all of the classical experiments in electricity and magnetism for himself. It was in his attempt to understand the nature of Oersted's recent discovery that electric currents produce magnetic effects that he discovered the phenomenon which was to become the basis for the electric motor. Faraday described his discovery in the letter to De la Rive as follows:

> I find all the usual attractions and repulsions of the magnetic needle by the conjunctive wire are deceptions, the motions being not attractions or repulsions, nor the result of any attractive or repulsive forces, but the result of a force in the wire, which, instead of bringing the pole of the needle nearer to or further from the wire, endeavours to make it move round in a never-ending circle and motion whilst the battery remains in action.[16]

That is, Faraday observed that the needle rotates rather than being pushed back and forth at right angles to the force. He went on to assert that the force is "simple and beautiful"; i.e., it is a *circular* force *surrounding* the current-carrying wire which causes either the wire or the magnet to rotate, rather than the complex attractive and repulsive distance forces described by Ampère.

Faraday's argument demonstrates both the primacy he attached to experimental observations and his lack of mathematical knowledge and sophistication. He observed that the needle rotates and assumed that to be the 'true' motion. However, the 'true' motion could be mathematically reconstructed in terms of central forces emanating from current elements in the wire, as Ampère had done, with the observed circular motion being simply the *resultant* of all the attractive and repulsive forces at work. Perhaps more importantly, though, Faraday's argument demonstrates his willingness to call the accepted Newtonian interpretation into question: He believed that a non-central force could be "simple" (i.e., irreducible). He did not take his analysis much further and, for the next ten years, wrote almost nothing about electricity and magnetism, not even in his diary. It was in 1831, in connection with his successful attempt to produce electric effects from magnetism, that he introduced the lines of force into his discussion of these actions.

Before turning to that discovery let me just add the following remarks to help elucidate Faraday's procedure. Throughout the

Researches, Faraday argues against the action-at-a-distance conception and for a new conception of electric and magnetic actions. His arguments against action at a distance attack what he takes to be essential components of that conception: (1) instantaneous, (2) straight line (i.e., along straight lines joining the centers of the bodies), and (3) not affected by the intervening medium. We can see that the argument given for "rotations" is against (2).

Faraday, of course, realized that a demonstration that electric and magnetic actions take time (against (1)) would be fatal, since that observation could not be reasoned away mathematically, but he was unable to devise an experiment to show it. That he firmly believed this is supported by a sealed letter he submitted to the Royal Society in 1832 (in order to claim priority for the idea):

> Certain of the results of the investigations which are embodied in the two papers entitled *Experimental Researches in Electricity*, lately read to the Royal Society: and the views arising, therefrom in connexion with other views and experiments, lead me to believe that magnetic action is progressive, and requires time; i.e. that when a magnet acts upon a distant magnet or piece of iron, the influencing cause (which I may for the moment call magnetism), proceeds gradually from the magnetic bodies and requires time for its transmission which will probably be found to be very sensible.
>
> I think also, that I see reason for supposing that electric induction (of tension) is also performed in a similar progressive time.
>
> I am inclined to compare the diffusion of magnetic forces from a magnetic pole to the vibrations on the surface of disturbed water, or those of air in the phaenomena of sound; i.e. I am inclined to think that the vibratory theory will apply to these phaenomena, as it does to sound and most probably to light.
>
> By analogy it may possibly apply to the phaenomena of induction of electricity of tension also.[17]

Although all of his arguments against parts (2) and (3) are at best inconclusive and at worse erroneous, they are at least explicit and clear. The discussions of his new conception of the actions are generally not very clear. He had to depend primarily on vague analogies with 'known' phenomena in order to articulate his conception. We can see such reasoning already in the above letter, where he likens the "progressive" actions to "vibrations" on water or in air.

3.3 Electromagnetic induction

Faraday's field concept developed over three periods of research: (1) research into and the discovery of electromagnetic induction; (2) research into electrochemical and electrostatic induction, from which came the discovery of the 'specific inductive capacitance' of dielectric media; and (3) research into magnetic induction, from which came the discoveries of the existence of 'diamagnetic' and 'magnecrystallic' media, the rotation of the plane of polarized light by magnetic action, and his demonstration of the essential dipolar nature of magnetism.

It was in the first period that Faraday introduced the lines of force into the description of the actions. As Oersted had shown, a constant electric current in a wire produces magnetic effects. Faraday had examined this phenomenon further, discovering his proposed "rotations." By symmetry, it was reasonable to expect a constant magnetic force to produce electric effects. For example, a conductor should have a current produced in it by a bar magnet. This had not been observed experimentally. It was Faraday's discovery that a *changing* magnetic force was needed.

His discovery came in two steps. First, he observed that the switching on and off of a current in a loop induced a current in a nearby conducting loop. He called this effect "volta-electric" induction. Second, given a permanent magnet and a nearby conducting loop, the motion of either produced a current, which died out as it came to rest. He called this effect "magneto-electric" induction. Faraday recognized the two effects to be the same and at first postulated a "new condition of matter" to explain why only a changing magnetic force induces a current:

> Whilst the wire is subject to either volta-electric induction or magneto-electric induction, it appears to be in a peculiar state: for it resists the formation of an electrical current in it, whereas, if in its common condition, such a current would be produced; and when left uninfluenced it has the power of originating a current which the wire does not possess under common circumstances. This electrical condition of matter has not hitherto been recognized, but it probably exerts a very important influence in many if not most of the phaenomena produced by currents of electricity. For reasons which will immediately appear (71), I have, after advising with several learned friends, ventured to designate it as the *electro-tonic* state.[18]

What is meant here is that the presence of a magnetic force *always* produces a state of tension within a conductor: the "electro-tonic state." The introduction or removal of the force builds up or releases the strain, thus allowing the production of a current:

> In all those cases where the helices or wires are advanced towards or taken from the magnet (50.55.), the direct or inverted current of induced electricity continues for the time occupied in the advance or recession: for the electro-tonic state is rising to a higher or falling to a lower degree during that time, and the change is accompanied by its corresponding evolution of electricity.[19]

The "electro-tonic state" provided a temporary solution to the problem of how *change* in a magnetic force could produce a current. The presence of a magnetic force, constant or changing, always produces a state of tension in a conductor. Just what this state of "tension" was posed a problem. Faraday seems to have considered it initially as a kind of 'polarization' of the particles of the conductor, but not ordinary electrical polarization. Partially because of his failure to detect the state experimentally, Faraday temporarily put aside the notion. However, as we will see, he was to reintroduce it in his work on dielectrics, as the explanation for the lateral repulsion of the lines of force. Finally, in his work on magnetic induction, the "electro-tonic state" was to become a tension in the lines of force themselves. The other reason he originally put aside the "new condition of matter" came from his attempt to deal with the problem of how a current could arise in a conductor *simply because of motion.* This attempt led him to introduce the concept of lines of force into the description of electric and magnetic actions.

It was well known that if iron filings are sprinkled around a magnet, they form discrete curves, originating at the poles and extending into the space surrounding the magnet. As the strength of the magnet increases, new lines develop within and grow outward, causing the whole system to expand outwardly (see Figure 1). For Faraday, this situation came to provide a graphic illustration of the *actual* transmission of electric and magnetic actions. The key question that led him to put the emphasis on the lines of force was:

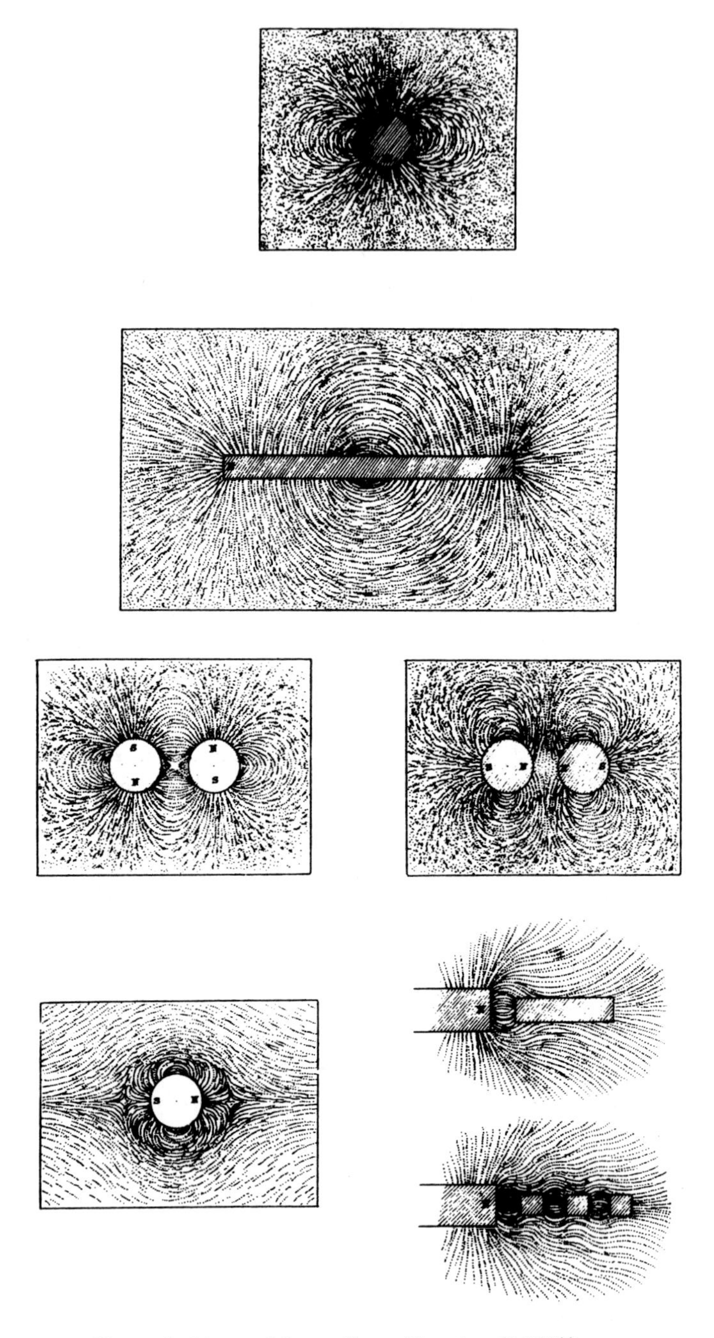

Figure 1. Lines of force (from Faraday (1831)).

> . . . whether it was essential or not that the moving part of the wire should, in cutting the magnetic curves, pass into positions of greater or lesser magnetic force; or whether, always intersecting curves of equal magnetic intensity, the mere motion was sufficient for the passage of electricity.[20]

Experiment showed the latter to be true. Motion through an area of constant magnetic force induces a current in a conductor, just as with an increase or decrease in the force. It is not necessary for the conducting wire to pass into greater or lesser areas of magnetic force, simply "cutting" the lines is sufficient to produce a current. Faraday even demonstrated that a quantitative relationship holds between the number of lines cut and the induced force.[21] He also extended this conception to include the case where there is no motion, i.e., the two loop, changing current case, by viewing the lines themselves as *moving out* from the loop carrying the initial current and passing over the second loop, thus being 'cut' by it; thereby producing a current. Additionally he discovered that it is possible to induce a current in the metal of a rotating magnet. He conceived of the current as being produced by the magnet *cutting its own lines of force*. From this he was led to claim that a "*singular independence* of the magnetism and the bar in which it resides is rendered evident."[22] He did not discuss the consequences of this claim, but I suspect that this phenomenon played a significant role in his thought. If the magnetism does not reside in the bar, where does it reside? Quite probably in the space surrounding the bar. He had expressed similar thoughts earlier, in an 1823 entry into his diary, in connection with the "rotations." There he noted that it "appears that the action does not regard the material substance at the centre but simply the current of electricity or the magnetic pole, *the substance merely giving locality to the power* [italics mine]."[23]

The rest of Faraday's *Researches* are concerned with working out and extending, by experiment and speculation, the hypothesis made here: that electromagnetic induction involves a relationship between the conductor and the lines of force. We must, however, be careful about how much of an initial speculation we attribute to him. It can be reasonably assumed at this juncture that he believed that: (1) induction is not an action at a distance, but a new action; (2) this new action requires further study, which should give a new understanding of electric and magnetic forces in general; and (3) the lines of force would quite probably be shown to be essential to the description of the actions. The speculation concerning the lines of force *guided* his future research.

How did Faraday conceive of the lines of force at the point of their introduction? There is significant ambiguity in what he said about them, and this ambiguity remained until his 28th series of researches, where he attempted to clarify the meaning he attached to them. The following remarks, made in discussing his observation that induction takes place by a change in the current of the source even when there is no motion between the two conductors involved, illustrate the ambiguity quite nicely:

> In the first experiments (10.13.) the inducing wire and that under induction were arranged at a fixed distance from eachother, and then an electric current sent through the former. In such cases *the magnetic curves themselves must be considered as moving* [italics mine] (if I may use the expression) across the wire under induction, from the moment they begin to be developed until the magnetic force of the current is at its outmost; expanding as it were from the wire outwards, and consequently being in the same relation to the fixed wire under induction, as if *it* moved in the opposite direction across them, or towards the wire carrying the current. Hence the first current induced in such cases was in the contrary direction to the principal current (17.235.). On breaking the battery contact, the *magnetic curves* (*which were mere expressions for arranged magnetic forces*) *may be conceived as contracting upon and returning* [italics mine] towards the failing electrical current, and therefore move in the opposite direction across the wire, and cause an opposite induced current to the first.[24]

Here the lines are seen as "moving," "expanding," "contracting," i.e., as behaving as real physical entities would; while at the same time they are qualified as being "mere expressions" for the arrangement of the forces. Faraday later called the latter "representative" lines of force and the former "physical" lines of force. He compounds this confusion in his later work by sometimes speaking of the physical lines as the *paths* of transmission of the force and sometimes as the *transmitters* of the force (i.e., as the vehicles of transmission). With electrostatic induction, he indicates the 'paths' conception; with magnetic induction, he indicates the 'transmitters' conception. The distinction is somewhat odd and will be discussed in more detail in what follows. The primary reason it occurs is that in the former case, he never felt he had been able to demonstrate the existence of the electrostatic lines of force as independent of the particles of dielectric media, while he felt he had been able to at least argue convincingly for the independent existence of the magnetic lines of force. Let us now turn to his conception of the lines of force in his second period of research.

3.4 Electrostatic induction

At the outset of the series of researches into electrostatic induction Faraday claimed that he was led to suspect that electrostatic induction was not a case of action at a distance by his work with electrolytes. There he had come to the view that the cause of electrochemical decomposition was to be found in forces *internal* to the medium.[25] He argued for electrostatic actions being transmitted through the intervening medium as well:

> . . . as the whole effect in the electrolyte appeared to be an action of the particles thrown into a peculiar or polarized state, I was led to suspect that common induction was in all cases an *action of contiguous particles* and that electrical action at a distance (i.e. ordinary inductive action) never occurs except through the influence of intervening matter.
>
> The respect which I entertain towards the names of Epinus, Cavendish, Poisson, and other most eminent men, all of whose theories I believe consider induction as action at a distance and in straight lines, long indisposed me to the view I have just stated; . . . it is only of late, and by degrees, that the extreme generality of the subject has urged me still further to extend my experiments and publish my views. At present I believe ordinary induction in all cases to be an action of contiguous particles consisting in a species of polarity, instead of being an action of either particles or masses at sensible distances; and if this be true, the distinction and the establishment of such a truth must be of the greatest consequence to our further progress in the investigation of the nature of electrical forces.[26]

Thus, Faraday conceived of electrostatic induction as a *polarization* phenomenon, consisting in the polarization of the "contiguous" particles of the intervening medium by the charges. That is, the charges affect each other through the action of the intervening medium, rather than directly. If he could demonstrate this experimentally, he felt it would provide him with an argument against component (3) of action at a distance. In the course of his electrostatic researches, Faraday attempted to work out a unified view of induction and conduction, which took the form of a 'wave' model of their interconversion.[27] He also attempted, unsuccessfully, to present a more unified view of electric and magnetic actions by demonstrating that electrostatic induction produces magnetic effects.[28]

What Faraday meant by "contiguous particles" is the key to understanding his conception of electrostatic induction and has thus been the subject of much controversy then and now.[29] About a year after he gave the above statement of his views, he added an explanatory note, which is more confusing than explanatory:

> The word *contiguous* is perhaps not the best that might have been used here and elsewhere; for as particles do not touch eachother it is not strictly correct. I was induced to employ it, because in its common acceptation it enabled me to state the theory plainly and with facility. By contiguous particles I mean those which are next.

Given this definition of "contiguous," he appears simply to be replacing action at what he called "sensible distances" with action at a distance between contiguous particles. Indeed, the confusion is compounded by his discussion of induction across a vacuum. In this case, "the next existing particle would be considered as the contiguous." In such cases, a particle could, e.g., "act at a distance of half an inch," since he does not mean that "there is *no* space between" contiguous particles.[30] Taken together, his statements about electrostatic actions appear inconsistent, and he was challenged by Hare to explain the inconsistency.[31] Faraday's response adds little to what he said in his *Researches*, since he maintained that he did not see any inconsistency! He stressed the point that the apparent inconsistency may arise out of Hare's failure to grasp the meaning *he* (Faraday) attached to the words:

> I feel that many of the words in the language of electrical science possess much meaning; and yet their interpretation by different philosophers often varies more or less, so that they do not carry exactly the same idea to the minds of different men: this often renders it difficult, when such words force themselves into use, to express with brevity as much as, and no more than, one really wishes to say.[32]

Faraday clearly saw that he was creating new meanings for the "words in the language of electrical science," but did not go on to give them explicitly. By way of clarification he noted that while the case of induction across a vacuum is not "ordinary" induction, but a "very hypothetical case," it should come under the "same principles of action." All "ordinary induction" is by polarization of contiguous particles of the dielectric medium. As an "illustration of [his] meaning," he offered the distinction between radiation and conduction of heat: The former travels large distances to its receiver, the latter passes from contiguous particles, yet both are assumed to be the same type of actions.[33]

So, in order to understand Faraday's new meaning, we must turn to his electrostatic researches, where he is supposed to have given it. Two distinct points are at issue here: Faraday's arguments against the action-at-a-distance conception of the actions and his explanations of his lines of force conception. His main arguments against action at a distance, in the electrostatic case, are based on two experimental observations: (1) transmission of the force is affected by the action of the intervening medium (which he considered to be molecular) and (2) the lines of force appear curved.

Faraday's most important discovery in this period was that, contrary to the accepted form of Coulomb's law, there is a difference in the inductive capacity of various dielectric media. This discovery required a revision of Coulomb's law to include the 'specific inductive capacity' of the medium. Faraday's hypothesis as to why there is a difference was that a state of polarization is produced among the particles of the medium, which affects the number of lines of force which can pass through it. However, he did not follow this out because he could not demonstrate the existence of the electrostatic lines of force independently of the particles of the medium.

Faraday was also aware that his was not the only possible interpretation. While the discovery posed a problem for the action-at-a-distance conception, it did not amount to a refutation of the view that bodies and charges act on each other directly, regardless of the intervening medium. The experimental results could be accounted for by maintaining that given, for example, two charges, though each charge directly affects the other, each charge also directly affects the medium, and the altered medium in turn affects the charges. The observed difference in force with different dielectric media can be explained as the result of the combination of all these *direct* effects. Faraday placed more importance on (2), i.e., he felt he could prove that the electrostatic lines of force represent the actual 'paths' of transmission of the action, even if he could not prove that they were the actual 'transmitters' of the actions. Though he offers the argument in many forms, the following quotation best demonstrates his line of reasoning:

> . . . induction [electrostatic] is exerted in lines of force which, though in many experiments they may be straight, are here curved more or less according to circumstances. I use the term *line of inductive force* merely as a temporary conventional mode of expressing the direction of the power in the cases of induction; and in the experiments with the hemisphere (1224), it is curious to see how, when certain lines have terminated on the under surface and edge of the metal, those which were before lateral to them *expand and open out from eachother*, some bending round and terminating their action on the upper surface of the hemisphere . . . . *All this appears to me to prove that the whole action is one of contiguous particles, related to eachother, not merely in the lines which they may be conceived to form through the dielectric, between the 'inductric' and 'inducteous' surfaces (1483), but in other lateral directions also* [italics mine]. It is this which gives an effect equivalent to a lateral repulsion, or expansion of the lines of force I have spoken of, and enables induction to turn a corner (1304). The power instead of being like that of gravity, which causes particles to act on eachother through straight lines [sic!], whatever other particles may be between them, is more analogous to that of a series of magnetic needles, or to the condition of the particles considered as forming the whole of a straight or curved magnet. So that in whatever way I view it, and with great suspicion of the influence of favourite notions over myself, I cannot perceive how the ordinary theory applied to explain induction can be a correct representation of that great electrical action.[34]

It is important to discuss this argument in detail now, since Faraday uses similar reasoning in the case of the magnetic lines of force as well. His argument is not totally mathematically naive. There is geometrical warrant for his opinion that if two bodies or charges act mutually along lines which are curved, then something in the intervening space could contribute to determining the line of action. If there are two objects, considered with 'point-like' properties only, then the only uniquely determined line of action between them (presuming the usual spatial symmetries) is a straight line. If the bodies act along some other line, it is reasonable to assume the existence of 'off-axis effects'. That is, single line geometrical confinement precludes a physical process in the intervening space, since the force would be of infinite strength; while transmission along other lines leaves open the possibility of processes occurring in the space (but does not, of course, establish that there actually are any). In the case of electrostatic induction, Faraday seemed to think that the action in curved lines shows the existence of some "power" responsible for the lateral repulsion of the lines of particles. He reintroduced his notion of "electro-tonic state" to represent this lateral tension, but just what it is is not any clearer here than when he had first introduced it in connection with his discovery of electromagnetic induction.

However, although Faraday could not "conceive how the ordinary theory . . . can be a correct representation" of the action, in fact it can. His argument that the experimental results showed that the actions are actually transmitted along curved lines rests on a confusion. He failed to make a distinction between the *resultant path* of the forces and the *actual path* of transmission of the forces involved. When this distinction is made, it is quite simple to explain the experimental observations from an action-at-a-distance point of view (once we make the above-mentioned modification of Coulomb's law). All the charges can be held to act directly upon each other along straight lines joining their centers (pairwise) and also on the particles of the intervening medium in the same way, with the affected medium then acting back on the charges. The summation of all these actions would align the particles of the medium in such a way that they would assume the shape of curved lines. However, in such a case the curved lines would not represent the *actual* paths of transmission, but only the *resultant* of all the straight-line actions involved.

As stated before, Faraday was to use the same line of argument with magnetic induction, but felt his case was even stronger there because the magnetic lines of force appeared to exist as conditions of "mere" space. The essential differences between the two cases lie in his contention that while electrostatic effects are polarization effects, magnetic effects are not. The significance of such a distinction will be discussed in detail shortly, after having made sense of his conception of electrostatic induction. In order to do so, let us first summarize the discussion of the curvature argument.

Because electrostatic induction, as conceived by Faraday, involves polarization of the particles of dielectric media, we are left with two possibilities: either (1) the force is actually transmitted along the observed curved lines or (2) the apparent lines of force simply represent the resultant forces on the particles, while the forces are actually transmitted directly. The observed phenomena do not present a decisive case for either hypothesis.[35] Faraday's hypothesis is (1), but his failure to see the possibility of (2) was in part the result of the confusion we just discussed.

But his failure to see (2) was perhaps also the result of the "influence of favourite notions" over his thinking, for, if we look further into his electrostatic researches, we see that the *curvature* of the lines played another important role in his conception of electric and magnetic actions. Faraday argued that "both *induction* and *conduction* appear to be the same in their principle of action."[36] His 11th to 14th series of researches are concerned primarily with formulating a relationship between static electricity and current electricity; between insulation and conduction; and between induction and conduction.[37] His conception of their connection appears to take the form of a 'wave' model of interconversion. As the particles of any medium become polarized, the lines of force expand outward to a maximum state of tension and thus obtain a maximum curvature. What the maximum is depends upon the particular medium and, e.g., would be significantly different for good insulating media as compared with good conducting media. The collapse of the lines would create a current in a conductor. The amount of curvature the lines obtain would represent the intensity of the force.[38] The lateral tension between the lines represents the "electro-tonic state." Faraday hoped to include magnetism in this unified picture, as a 'vibratory' effect, but was unable to show that static electricity produces the required magnetic effects.[39]

It should be noted that Faraday did not present this conception as succinctly as I have stated it. However, given the series of researches in question and his 1846 "Thoughts on ray vibrations," in which he attempted to incorporate light into the picture (also as 'vibrations' of the lines of force), one can construct the simplified description presented here.[40] Given his conception of induction and conduction, it is clear that Faraday would not have accepted the action-at-a-distance explanation of the curvature as fatal to his view. It should also be clear to the reader that he was giving new meanings to the "words of electrical science," as he suggested to Hare.

Given our present understanding of what is in those sections of his researches to which he refers Hare, we can return to the question of what he meant by saying that electrostatic actions take place by the action of "contiguous particles." We can now see how the successive polarization of the particles of dielectric media (conducting media as well, given the above) would produce the observed effects: the less polarizable the medium, the less tension produced in the medium, resulting in a greater inductive capacitance and vice versa. What remains unclear, though, is *how* one particle affects the next. Surely Faraday wanted a more significant contrast of views than simply trading action at a distance at large distances for action at a distance at small distances. Even if we accept the argument given by Gooding that Faraday changed the conception of the actions significantly by allowing a particle to affect only the *next* successive one, it is still action at a distance.[41] Faraday also had a new conception of *how* the action takes place. The way to understand his new conception is to incorporate his conception of particles into the analysis. Unfortunately he only expressed his views concerning particles in the later "Speculation . . . ."[42] However, I think it is a legitimate move to incorporate those thoughts at this point, since he said that "certain facts from electrical conduction and chemical combination" made him think about particles in the way he presented them in this later paper.[43] Faraday considered his conception of particles to be among the most speculative of his views, since the experimental evidence he had to offer in support of it was negligible, which is why he was reluctant to work it into his researches. But, as he said in his notes to the lectures he gave on the subject, there is a "time to speculate as well as to refrain," and he realized that a statement of his speculations on the subject was necessary to produce a coherent view of the actions.[44]

In his "Speculations . . . ," Faraday argued that a paradox arises if we assume the standard conception of matter as composed of atoms separated by empty space: Space must be both a conductor and an insulator. He proposed instead that there is no empty space. Rather, material particles are point centers of converging lines of force – an idea similar to, though not identical with, that of Boscovich.[45] Every 'particle' is thus connected with every other, through the lines of force, and the sphere of its action extends throughout "the whole of the solar system." He maintained that this was actually a simpler way to think of matter because all we really are familiar with are the forces in nature and not something in addition to them called 'matter'. Thus, force would be the only substance, all forces would be interconvertible, force would be conserved, and there would be no need for an aether to act as a medium of transmission.[46] Given this conception, "contiguous particles" would act upon one another through their mutual lines of force. If we insert this picture into what Faraday said in his research articles, the inconsistency Hare noted disappears.

Faraday was hesitant to express his view concerning particles not only because he lacked experimental evidence for it, but also because he had no experimental evidence for the existence of the lines of force at all. All he had shown in the second period of research is that the particles of dielectric media assume a certain configuration when subject to electrical force. We will now turn to his third period of research, primarily on magnetic induction, and examine the evidence he presented there for the view that the lines of force exist independently of the particles of magnetic media – as states of "mere space."

3.5 Magnetic induction

The final period of Faraday's research began with his discovery that the action of a magnet affects the plane of polarization of light. This discovery provided the long sought after experimental evidence for the connection between light and electricity and magnetism. It also reinforced his views on the unity and interconversion of all forces.[47] He maintained that the polarized light "illuminated" the lines of magnetic force. He did not mean, however, that we could see the lines directly, but that by "[e]mploying a ray of light, we can tell, *by the eye*, the direction of the magnetic lines through a body; and by the alteration of the ray and its optical effect on the eye, can see the course of the lines just as we can see the course of a thread of glass, or any other transparent substance, rendered visible by the light."[48]

Continuing his research into magnetic actions, Faraday first tried to explain magnetic induction in a way similar to his explanation of electrostatic induction, but encountered problems. While "ordinary" magnetic media ('paramagnetic') seemed to fit the polarization model, Faraday discovered another type of magnetic medium, which he called "diamagnetic," whose action left him with a puzzle. The tendency of diamagnetic media is to move from stronger to weaker points of force and, when in a uniform field, they align themselves perpendicular to the lines of force, rather than parallel (as with paramagnetic).

Faraday first hypothesized that "a contrary state might be produced in ordinary magnetic substances and in diamagnetic."[49] In each case polarization would take place in the particles of the medium, and the axis of polarization would be parallel to the resultant of the magnetic force passing through it. With paramagnetism, contrary poles would face the inducing magnet, while with diamagnetism, the opposite would occur. Thus, in the former case we have attraction, while in the latter we have repulsion. There is, however, a major problem with this explanation. If we have two parallel loops of wire and create a magnetic force by starting a current in one, a current will be produced in the other having the opposite direction, thus producing a magnetic force with opposite orientation to the first. The two will have like poles facing and will repel each other. This situation corresponds to diamagnetism. But, it is not possible to account for paramagnetism in this way. Given the two loops, the induction which takes place in a paramagnetic body would have to correspond to a current in the secondary circuit in the same direction as the primary, instead of the opposite, which is not what happens.

Faraday's discovery of what he called "magnecrystallic" media at first complicated the situation further, since these substances indicate a third category of magnetic action. They display both diamagnetic and paramagnetic properties depending upon their orientation. The influence of a magnetic force on such media causes neither attraction nor repulsion, thus the induced force is not a polar force. It was in connection with these substances that he introduced the idea that the crystallized state might allow the "conduction" of the lines of force more easily in one direction than another.

Extending this idea to paramagnetism and diamagnetism enabled Faraday to unify the three seemingly distinct types of phenomena in terms of one conception of the action. Different media would have different degrees of "conductive power." He argued that this conception of the actions was a fruitful way to view them because of its unifying power, even though he was not able to establish how such conduction would take place:

> . . . if bodies possess different degrees of *conducting power* for magnetism, that difference may account for all the phaenomena, and, further, if such an idea be considered; it may assist in developing the nature of magnetic forces . . . . As yet, however, I only state the case hypothetically, and use the phrase *conductive power* as a general expression of the capability bodies may possess of effecting the transmission of magnetic force; implying nothing as to how the process of conduction is carried on. Thus limited in sense, the phrase may be very useful, enabling us to take for a time, a connected, consistent general view of a large class of phaenomena; may serve as a standard of meaning amongst them, and yet need not involve any error, inasmuch as whatever may be the principles and condition of conduction, the phaenomena dependent on it must consist among themselves.[50]

Thus, the different magnetic media could be viewed as allowing the lines of force to pass through them with varying degrees of ease. What the process of "conduction" itself involved was left open. It could be a polarization effect, or some other unknown process. However, the polarization would not be of the particles of the magnetic medium. In fact, Faraday hoped to establish that the polarity involved in magnetism resides *in the lines of magnetic force themselves.* The most conclusive piece of evidence, for him, was his demonstration of the essential dipolar nature of magnetism. Faraday showed that no matter how small, a "true magnetic system" is always dipolar. This convinced him that the lines of force constitute an essential part of a magnetic system, and magnetic action, in that *they* relate the observed polarities. Any action should take place through them, rather than by two separate poles acting at a distance on one another and on the medium. The dipolar nature of magnetism also figured in a key way in his arguments for the curvedness of the lines of force and for their existence as states of space.

As we have seen, Faraday was not very clear about what he meant by "lines of force." He acknowledged this himself and, with his 28th series of researches, he set out to clarify and state his views explicitly by attempting anew to explain the results of his researches in terms of the lines-of-force concept. He argued that, with respect to magnetism in particular, although the more widely accepted conception of the actions could account for all the phenomena and give the correct results, his conception had a greater potential for *explanation* of the actions:

> Now it appears to me that these lines may be employed with great advantage to represent the nature, condition, direction and comparative amount of the magnetic forces; and that in many cases they have, to the physical reasoner at least, a superiority over that method which represents the forces as concentrated in centres of action, such as the poles of magnets or needles; or some other methods, as, for instance, that which considers north or south magnetisms as fluids diffused over the ends or amongst the particles of a bar. No doubt, any of these methods which does not assume too much, will, with a faithful application give true results; and so they all ought to give the same results as far as they can respectively be applied. But some may, by their very nature, be applicable to a far greater extent, and give far more varied results than others. For just as either geometry or analysis may be employed to solve correctly a particular problem, though one has far more power and capability, generally speaking than the other; or just as either the idea of the reflexion of images, or that of the reverberation of sounds may be used to represent certain physical forces and conditions; so may the idea of the attractions and repulsions of centres, or that of the disposition of magnetic fluids, or that of lines of force, be applied in the consideration of magnetic phaenomena. It is the occasional and more frequent use of the latter which I at present wish to advocate.[51]

The lines-of-force conception of the actions has "far more power and capability" since it potentially provides a means for unifying all electric and magnetic actions; indeed, it is possible that *all* forces of nature could be unified under this conception.

At this point Faraday explicitly divided the lines-of-force concept into the two possibilities we have discussed: (1) They are "representative" of the intensity and direction of the force at a point and (2) they are the "physical mode" of transmission of the force (but still either the 'paths' or the 'vehicles' of transmission). He maintained that (1) was a nonspeculative, experimentally supported interpretation of what he meant in the *Researches*; while (2) was his "favourite notion":

> . . . I have sometimes used the term *lines of force* so vaguely, as to leave the reader doubtful whether I intended it as a merely representative idea of the forces, or a description of the path along which the power was continuously exerted [i.e., 'paths' *or* 'vehicles' in our analysis].
>
> What I have said at the beginning of this paper (3075) will render that matter clear. I have as yet found no reason to wish any part of those papers altered, except these doubtful expressions: but that will be rectified if it be understood, that, wherever the expression *line of force* is taken simply to represent the disposition of the forces, it shall have the fullness of that meaning; but that wherever it may seem to represent the idea of the *physical mode* of transmission of the force it expresses in that respect the opinion to which I incline at present. The opinion may be erroneous, and yet *all* that relates or refers to the disposition of the force will remain the same.[52]

The conception of the lines of force as "merely representative" of the intensity and direction of the force does not in itself distinguish Faraday's view from action at a distance. It is a vectorial notion, which can be used in the action-at-a-distance analysis as well. What Faraday wished to do here was to separate the results of his experimental researches from his speculations about what they indicated concerning electric and magnetic actions. What sets him apart is that he saw the "representative" notion as *neutral* between his more speculative view and the action-at-a-distance conception, which, he implied, was no less speculative. The "representative" notion is neutral and accessible to experimental investigation. It also protects his *Researches* from the objections which can be raised against his speculative views. In contrast to this stands the speculative "physical" notion that the lines of force actually represent some state of, or are due to some process in, the intervening space and that this state or process may be of fundamental importance for the theory of electricity and magnetism.

This speculation, in its general form, was to be vindicated by Maxwell's theory and its subsequent confirmation. Faraday's own particular version of this speculation considered the lines of force as 'lines of transmission' of force, which is ambiguous between 'paths' of transmission and 'vehicles' of transmission.[53]

Faraday's attempt to maintain a separation between fruitful ways of understanding the experimental results and speculations about the "physical cause of the phaenomena" led him to publish his speculations in *Philosophical Magazine*, rather than in *Philosophical Transactions*, where his researches appeared! In "On the physical character of lines of magnetic force," he directly tackled the question of whether the lines of magnetic force are physical entities.[54] There he offered what he considered to be the strongest arguments in favor of the conception that the lines of force are the lines of transmission of magnetic actions. It is clear from the arguments he presented that he intended the 'vehicles' conception. He distinctly contrasted the electrostatic case, where the lines exist "by a succession of particles," with the magnetic case, where the lines exist (if they do) "by the condition of space free from particles."[55] As was noted before, for him to make this distinction was rather odd. Given his conception of particles, it could not be a distinction *in kind.* Rather, it strikes me as a distinction in *evidence*, i.e., Faraday felt he had more conclusive evidence for the existence of the lines in the magnetic case than in the electrostatic case, whereas he had none for his conception of particles. Let us now turn to the nature of the evidence he presented.

He divided the concept of action at a distance into the three components we noted earlier: (1) instantaneous, (2) straight line, and (3) not affected by matter in the intervening space. As in the electrostatic case, his argument is that (2) and (3) are not true of magnetic actions. He recognized that (1) is the only component which would provide conclusive evidence against action at a distance, but he was unable to show that magnetic actions take time. Such a demonstration would provide the crucial evidence that something was going on in the intervening space, i.e., that a "physical agency existed in the course of the lines of force."[56] Lacking this crucial piece of evidence, Faraday attempted to build his case on what he considered to be persuasive, albeit inconclusive, evidence.

Faraday compared magnetic action with the 'known' phenomena of gravitation and wave radiation. Though seemingly instantaneous, magnetic actions appear to be transmitted "by physical lines through the intervening space," (unlike gravitation) but not as a wave which "once produced has an existence independent of either its source or termination" (like radiation). They are more akin to electrical actions in which the "propagating process has intermediate existence like a ray, but at the same time depends upon both extremities of the lines of force, or upon conditions (as in the connected voltaic pile) equivalent to such extremities."[57] The major question is to what type of electric actions are they more akin? Is the process static like that of electric induction or dynamic like electrical conduction? Faraday concluded that the lines are states of static strain or tension. He again appealed to the notion of the "electro-tonic state" as a way of conceiving of this strain. However, the "electro-tonic state" seems to have changed its character in an important way. Rather than being a "new condition of matter," "such a state would coincide and become identified with that which would constitute the physical lines of magnetic force."[58] So, the "electro-tonic state" has become a way of characterizing the state of strain *in the lines of force themselves.* Since the lines of magnetic force are not composed of polarized particles, the "electrotonic state" is *dissociated from matter* and is "a condition of space free from material particles," as Faraday had said of the lines.

Given the lack of experimental evidence that magnetic actions take time, the bulk of Faraday's argument for the existence of the lines as conditions of space inevitably rests on a consideration of parts (2) and (3) of the notion of action at a distance. We have already examined similar arguments in the case of electrostatic induction. With (3), the observed differences in force transmitted through different magnetic media can be explained as the result of the summation of all the distance forces involved.

Faraday took the denial of (2) as the most crucial and conclusive point in his argument. He maintained that if the lines of force are curved, they are very likely to have physical existence:

> If an action in curved lines or direction could be proved to exist in the case of the lines of magnetic force, it also would prove their existence external to the magnet on which they might depend . . . .[59]

> . . . I cannot conceive curved lines of force without the conditions of a physical existence in that intermediate space.[60]

> To acknowledge the action in curved lines, seems to me to imply at once that the lines have a physical existence. It may be a vibration of the hypothetical aether or a state of tension of that aether equivalent to either a dynamic or static condition; or it may be some *other state*, which though difficult to conceive, may be equally distinct from the supposed non-existence of the lines of gravitational force and the independent and separated existence of the lines of radiant force.[61]

It has already been shown in the electrostatic case that there is geometrical warrant to support the claim that if the action is other than straight-line, it is reasonable to assume the existence of 'off-axis effects' in the intervening space. We also saw that by making the distinction between the resultant path and the actual path of transmission of force, the warrant disappears in the electrostatic case. However, Faraday's argument needs to be reconsidered here, since he held that magnetic induction does not involve polarization of the particles of magnetic media.

Reiterating what has been said thus far, there are two possible meanings to the claim that the action is in curved lines: (a) the 'paths of transmission' meaning: Here the curved lines would be composed of polarized particles and the force would be transmitted from particle to particle, the lines having no existence apart from the particles of the medium, and (b) the 'vehicles of transmission' meaning: Here the lines of force themselves would exist and would be the transmitters of the force, the particles aligning themselves along them. With meaning (b), the curvature would be the result of the amount of tension in the medium.

The most Faraday could have hoped to establish in the electrostatic case was (a). However, given (a), it is possible to explain the apparent curvature from an action-at-a-distance point of view as the resultant effect of all the direct actions involved, as we have seen. Even in this case, though, Faraday's preferred meaning seems to be (b) if we consider his 'wave' model of interconversion of electric induction and conduction.

In his discussion of magnetic induction, Faraday made it unmistakably clear that he intended meaning (b). Since magnetic induction does not produce polarization in magnetic media, he felt that he could make a strong case for the claim that the curvature resides in the lines of force themselves. What this would mean, for example, is that when we observe iron filings aligning themselves around a magnet, they are aligning themselves along the already present lines. However, the case is not as strong as Faraday believed. When we consider what experimental evidence there is for the existence of the lines of magnetic force, we always require reference to at least one other body, such as the filings, to detect them. Faraday was aware of the "summation of forces" objection and tried to "trace" the curvature of the lines by probing the area around the magnet with a conducting wire to see what orientations produced a current. However, this means of 'detecting' the lines also involves reference to another object besides the magnetic source! The situation is thus always turned into at least a three body problem and the geometrical warrant applies only to the two body case.

However, as we have noted, there is one additional piece of support for his claims in the magnetic case: Magnetism is dipolar. This 'fact' led him to hypothesize that the lines of magnetic force are always there and are always curved because they are what relate the "external polarities" of the magnet to each other. He had shown experimentally that the lines of force do not terminate at the poles, but pass through the magnet, relating the polarities in *closed curves.* From this he concluded that magnetic actions essentially involve interaction with the curved lines of force:

> I conceive that when a magnet is in free space, there is such a medium (magnetically speaking) around it . . . . What that surrounding magnetic medium, deprived of all material substance may be, I cannot tell, perhaps the aether. I incline to consider this outer medium as *essential* to the magnet; that it is that which relates the external polarities to eachother by curved lines of power; and that these must be so related as a matter of necessity . . . .
>
> . . . without this external mutually related condition of the poles, or a related condition of them to other poles sustained and rendered possible in like manner a magnet could not exist . . . .[62]

Also:

> In this view of a magnet, the medium of space around it is as essential as the magnet itself being part of the true and complete magnetic system.[63]

So, in Faraday's opinion, the lines of magnetic force do not just come into existence when we bring in a test object, but must always be there to relate the polarities of the magnet. However, given the impossibility of establishing this experimentally (we always need another object), it remained *his* opinion. It would take the mathematical formulation of the field theory of electric and magnetic actions to make his opinion a viable competitor to the action-at-a-distance view.

Faraday's conception of the lines of magnetic force as a "condition of space" is unquestionably a field concept. What he meant by a "condition of space" remained unclear. As we will discuss shortly, such a condition might involve an aether, but need not. However, it should be stressed that the concept of field does not enter *essentially* into the description of electric and magnetic actions at this point. Since there was no proof that the actions are not instantaneous, the case remained similar to that of Newtonian gravitation. Gravitational actions, in Newtonian theory can be formulated in terms of 'potential fields', which allow the derivation of the force which *would* act on a test particle if it *were* to be placed in the space surrounding the body. The same is true here. The concept of field is not necessary to the physical description of the phenomena since such fields are instantaneous and are completely determined by the arrangement of the masses and charges. A 'potential field' formulation of the lines-of-force concept would be simply a device for calculation, while the description of the phenomena could still be given in action-at-a-distance terms, as with gravitation.

Such a mathematical argument would, however, have counted very little with Faraday himself. Regardless of what mathematical formulations were possible, the physical data indicated to him that another *description* of the phenomena was at least possible, and the truth of that description seemed to him very probable. His reaction against the mathematical argument would not be simply the result of his mathematical ignorance. His approach to scientific inquiry placed the emphasis on experiment rather than on mathematical argumentation. As he said as early as 1831, "experiment need not quail before mathematics, but is quite competent to rival it in discovery."[64] He did see that a new mathematical formulation might be needed, but, in his "On some points of magnetic philosophy" of 1854, he indicated that perhaps, in the final analysis, mathematics would not be able to settle the issue.[65] He was impressed, though, by Thomson's work which showed the mathematical equivalence of various ways of representing magnetic actions. He was even more impressed with Thomson's analogy between heat flow and current flow, which are physical processes, and electric and magnetic lines of force. He felt that the analogy between the lines of force and known physical processes provided evidence for their being physical processes as well.[66] Finally, his reluctance to accept mathematical arguments as providing the final word on the subject stemmed from his belief that understanding the nature of electric, magnetic, and gravitational actions would require possession of a conception of *how* these actions take place in addition to any mathematical description.[67] The lines-of-force conception provided, at the very least, a powerful visual image of the transmission of electric and magnetic actions. The question of what kind of physical processes the lines of force could be remained unanswered for Faraday.

3.6 The aether

Before concluding this chapter, we need to address the question of what role the aether played in Faraday's concept of field. The question arises in connection with the problem of what kind of processes the lines of force could be. Given Faraday's conception of particles as point centers of converging lines of force, there would seem to be no need for an aether. The lines of force would be the only substances, and all interaction would take place through them. Thus, Faraday's "favourite notion" would have no need of an aether. However, his conception of particles was pure speculation, lacking in experimental support. When viewed in their entirety, his *Researches* were inconclusive with respect to the existence or nonexistence of any aether – even the light aether.[68] Given this, he remained uncommitted as to whether the lines represent some state of, or process in, an aether. His indecisiveness is best illustrated by the following:

> Whether it [the hypothesis of physical lines of force] of necessity requires matter for its substantiation will depend upon what is understood by matter. If it is to be confined to ponderable or gravitating substances, then matter is not essential to the physical lines of magnetic force any more than to a ray of light or heat; but if in the assumption of an aether we admit it to be a species of matter, then the lines of force may depend upon some function of it. Experimentally mere space is magnetic; but then the idea of such mere space must include that of the aether, when one is talking on that belief; or if hereafter any other conception of the state or condition of space rise up, it must be admitted into the view of that, which just now in relation to experiment is called mere space. On the other hand, it is, I think, an ascertainable fact, that ponderable matter is not essential to the existence of physical lines of magnetic force.[69]

Thus, Faraday saw no problem with the conception of the physical lines of force as conditions of "mere space" itself. If the aether exists, it is definitely not ponderable matter. He saw no objection to extending the conception of "mere space" to include such an aether. However, given all that we have seen about his views, it is quite possible that such an aether would be one to which Newton's equations would not apply. In any event, Faraday did not see the notion of any aether as essential to his argument for the physical existence of the lines of magnetic force. He felt that the experimental evidence supported their existence, whereas it was neutral with respect to the existence of an aether. Faraday's successors, Maxwell in particular, interpreted him as an aether theorist. (Maxwell always chose those quotations which made Faraday appear as though he believed in the aether.) This was a quite natural result of Faraday's own indecisiveness and of his reticence and inability to spell out fully his new conception.

3.7 Summary: Faraday's concept of field

We have seen that Faraday conceived of electric and magnetic actions as transmitted continuously across space through the lines of force. The two aspects of his lines-of-force concept can be summarized by Table 2.

His neutral conception protected his *Researches* from possible criticisms of his field conception. In contrast with the neutral conception he speculated that the lines are states of stress and

Table 2. Two aspects of Faraday's lines-of-force concept.

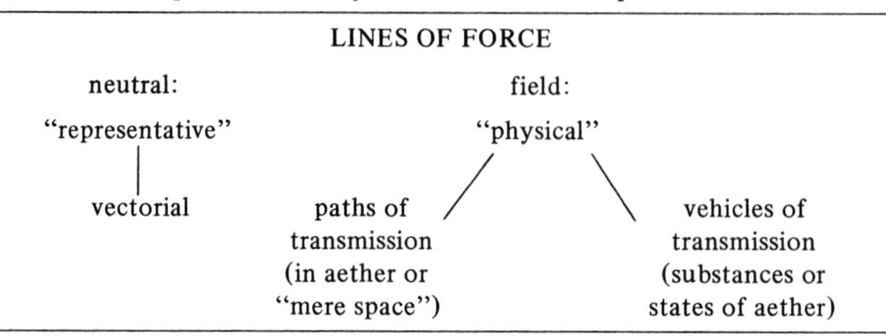

strain in space, which may or may not be filled with an aether. That the lines are substances existing in "mere space" seems to have been Faraday's preferred view and was an essential part of his 'grand unified' conception of the unity and interconversion of all the forces of nature. His more cautious hope was that, at the very least, the lines of force would prove essential to the description of electric and magnetic actions.

It should be stressed once more that Faraday was quite aware of the speculative nature of this conception. This is precisely why he felt that in reporting on his experiments he should not go beyond the neutral conception. Yet, he considered such speculations to be an integral part of the development of a new conception of physical actions. He argued that they should be given serious consideration, as, fortunately, was to be done by Maxwell and Thomson. What Faraday, himself, had to say about the role of speculation in science is instructive as well as prophetic:

> It is not to be supposed for a moment that speculations of this kind are useless, or necessarily hurtful, in natural philosophy. They should ever be held as doubtful, and liable to error and change; but they are wonderful aids in the hands of the experimentalist and mathematician. For not only are they useful in rendering the vague idea more clear for the time, giving it something like a definite shape, that it may be submitted to experiment and calculation; but they lead on, by deduction and correction, to the discovery of new phaenomena, and so cause an increase and advance of real physical truth, which, unlike the hypothesis that led to it, becomes fundamental knowledge, not subject to change.[70]

James Clerk Maxwell (1831–1879)

CHAPTER 4

Maxwell's 'Newtonian aether-field'

4.1 The "representative lines of force"

Maxwell took Faraday's conception of electrostatic and magnetic actions as involving continuous transmission of force, rather than the action-at-a-distance conception, as the starting point for the formulation of his own conception. He first attempted to give mathematical form to the "representative lines of force" and then to the "physical lines of force" of Faraday, while at the same time trying to incorporate electrodynamical theory into the framework of Newtonian mechanics. He felt this incorporation could be achieved by considering the lines of force to be states of a mechanical aether, i.e., one which would obey Newton's laws. Thus, he only focused on one aspect of Faraday's conception: that the line of force could be some state of a material, though nonponderable, medium.

Maxwell acknowledged his indebtedness to Faraday in many places. The clearest statement to this effect is that in a letter to Faraday in 1861, in which Maxwell conveyed the main ideas of his second major paper on electrodynamics:

> When I began to study electricity mathematically I avoided all the old traditions about forces acting at a distance, and after reading your papers as a first step to right thinking, I read the others, interpreting as I went on, but never allowing myself to explain anything by these forces. It is because I put off reading about electricity until I could do it without prejudice that I think I have been able to get hold of some of your ideas, such as the electro-tonic state, action of contiguous parts, etc., and my chief object in writing you is to ascertain if I have got the same ideas which led you to see your way into things, or whether I have no right to call my notions by your names.[71]

There is no evidence that Faraday replied, and it is most likely that he did not because of the advanced state of his illness. Given our analysis of Faraday, we can see that, by ignoring Faraday's favorite speculation, Maxwell did not "get hold" of all of the aspects of his field concept. However, I think it is fair to say that he did "get hold" of a significant aspect of Faraday's concept and, in attempting to develop that aspect mathematically, furthered the development of the concept of field. In what follows, I will be discussing but a small portion of Maxwell's contributions to electrodynamics. I will focus on those which relate most specifically to the formation of his electromagnetic field concept.

In discussing his thoughts in letters to William Thomson during the writing of his first paper on the subject, "On Faraday's lines of force," he stated that he had "appropriated" Faraday's view of " 'magnetic polarity' as a property of a 'magnetic field' or space" and "the general notion of lines of force and *conducting* power."[72] Additionally, he borrowed Thomson's method of analogy, which will be discussed shortly, and Ampère's theory of closed circuits, which he wanted to formulate in Faraday's terms.[73] With these in hand, he gave a mathematical formulation of the representative lines-of-force conception. The analogy he used was that between the intensity and direction of a line of force at a point and the flow of an incompressible fluid through a fine tube of variable section. What Maxwell's analysis provides is a vector representation of the lines of force in terms of the velocity field of a fluid.[74] In his analysis, he replaced Faraday's relationship between the number of lines cut and the strength of the force with a continuous measure: The tubes can be arranged in such a way that there are no spaces, so the fluid fills all space. The specific details of this analogy are not needed for our purposes. I will simply give the broad outlines of the paper and state those results he was to make use of in his later work.

After formulating the analogy, Maxwell applied it to static electricity, galvanic currents, permanent magnetism, and magnetic induction. In the second part of the paper, he applied it to electromagnetic induction and to Faraday's "electro-tonic state." One important result is that the analogy led him to distinguish between 'forces' ('intensity'; usually, but not necessarily a vector) and 'fluxes' ('quantity'; always vector) – a distinction that would remain throughout his work. Finally, he provided an "electrotonic function" as a representation of "electro-tonic state." This function is clearly what we now call the 'vector potential'. He derived the equations relating the magnetic intensity to this function and the induced electromotive force to changes in it, both of which he was to use in his next paper. However, Maxwell was dissatisfied with his analysis of Faraday's notion, because he could not find an interpretation of it in terms of the analogy. This was to be achieved with the analogy used in the next paper.

Maxwell wished the analysis in this first paper to be viewed as purely mathematical and as making no claims about the physical nature of the actions involved: i.e., it is only a descriptive analysis of the lines of force. The analysis, thus, offers no *argument* against the action-at-a-distance conception. In fact, the vectorial representation provided by Maxwell can be used with that conception of the forces as well. The difference comes in with his formulation of electromagnetic induction in terms of the "electrotonic function." He opposed this formulation, which he claimed does not contain "even a shadow of a true theory; in fact its chief merit as a temporary instrument of research is that it does not, even in appearance, *account for* anything," to Weber's "elegant" formulation of the action-at-a-distance conception of electromagnetism, which does involve a physical theory.[75] In response to expected challenges concerning the value of such an analysis, he maintained:

> Here then is a really physical theory [Weber's] satisfying the required conditions perhaps better than any yet invented; and put forth by a philosopher whose experimental researches form an ample foundation for his mathematical investigations. What is the use then of imagining an electro-tonic state of which we have no distinctly physical conception, instead of a formula of attraction which we can so readily understand? I would answer, that it is a good thing to have two ways of looking at a subject, and to admit that there *are* two ways of looking at it. Besides, I do not think that we have any right at present to understand the action of electricity, and I hold that the chief merit of a temporary theory is, that it shall guide experiment, without impeding the progress of the true theory when it appears.[76]

Thus, Maxwell saw the development of his formulation as performing two functions. First, it provided an alternative theory to guide experiments. Second, it demonstrated the existence of two ways of conceptualizing the same phenomena. Thus, one should keep an open mind about the subject. Maxwell's position is, then, like Faraday's in that he held both views to be speculations. Objectively, he had more reason to do so than Faraday. Although Weber had recently provided the long-awaited unified mathematical formulation of electricity (current and static) and magnetism, his formulation was not strictly Newtonian in that the relationship between the forces depends upon the *relative* velocity and acceleration of the charged particles. Also, at the time of Maxwell's first paper, it was thought that Helmholtz had shown that Weber's formulation violates the newly formulated principle of conservation of energy. (He was later shown to be wrong.)

4.2 The method of "physical analogy"

Before discussing the second paper, in which Maxwell first formulated the equations of the electromagnetic field, we need to discuss his method of "physical analogy," which he so successfully exploited in that paper. As we have seen already, Maxwell borrowed Faraday's conception of the continuous transmission of electric and magnetic forces as the starting point of his own work. Equally important to his contribution to the development of the field concept is that he borrowed William Thomson's method of analysis: "physical analogy."[77] With his analogies between heat and electrostatics and between light and vibrations of an elastic medium, Thomson had shown that it is possible to transfer the solution of a mathematical problem from one branch of physics to another; from an established branch of science to a newly developing one. The method involves using a "resemblance in form," i.e., an isomorphism, between the laws of different phenomena as a means of obtaining "physical ideas without adopting a physical theory."[78] Maxwell was greatly impressed with the power of Thomson's analogies, in particular, with their ability to provide new physical insights into the phenomena in question. While he was working on the lines-of-force paper, he asked Thomson if he had "patented that notion with all its applications? for I intend to borrow it for a season, without mentioning anything about heat (except of course historically) but applying it in a somewhat different way to a more general case to which the laws of heat do not apply."[79] Maxwell not only "borrowed" the notion, he, in typical Maxwell fashion, elaborated on it and its significance in such a way that he made the notion his own.

In the development of a new scientific conception, Maxwell argued:

> The first process therefore in the study of the science must be one of simplification and reduction of the results of previous investigation to a form in which the mind can grasp them. The results of this simplification may take the form of a purely mathematical formula or of a physical hypothesis. In the first case we entirely loose sight of the phenomena to be explained; and though we may trace out consequences of given laws, we never obtain more extended views of the connexions of the subject. If, on the other hand, we adopt a physical hypothesis, we see the phenomena only through a medium and are liable to that blindness to facts and rashness in assumption which a partial explanation encourages. We must therefore discover some method of investigation, which allows the mind at every step to lay hold of a clear physical conception, without being committed to any theory founded on the physical science from which that conception is borrowed, so that it is neither drawn aside from the subject in pursuit of analytical subtleties, nor carried beyond the truth by a favourite hypothesis.[80]

Thus, Maxwell saw the method, which he called "physical analogy," as a middle way between the sterility of a purely mathematical analysis and the excesses of speculative hypotheses. "Physical analogy" is "that partial similarity between the laws of one science and those of another which makes each of them illustrate the other."[81] That is, it provides a way of graphically exploring the possible physical implications of an isomorphism between the laws of different phenomena, without making any actual physical hypothesis; it provides a mathematical formalism plus a concrete, visual image to apply to the new phenomena.[82] Such exploration may, and hopefully will, lead to new physical hypotheses, which can be tested by experiment. The physical analogy used in the first paper had not led to any physical hypothesis. However, it had shown that an alternative representation of the forces was possible. The various physical analogies used in the second paper did lead to physical hypotheses: namely, that there is a time delay in the transmission of electric and magnetic actions and, though not made explicit until the third paper, that light is an electromagnetic phenomenon.

An analogy, itself, of course provides only partial solutions to the problems, and any analogy will be modified in the attempt to fit it to the specific problems in question and to complete the solution. Also, choice of a particular physical analogy produces constraints on the solutions as well as on the choice of further analogies. However, it should be stressed that Maxwell repeatedly emphasized the "provisional and temporary" character of physical analogies. A physical analogy does not contain "even a shadow of a true theory; in fact its chief merit as a temporary instrument of research is that it does not, even in appearance *account* for anything."[83] That is, it does not tell us how electric and magnetic actions *actually* are transmitted. It functions simply as a *heuristic device* for exploring the unknown phenomena. In the first paper, the fluid flow analogy had aided the mathematical formulation of the representative lines-of-force conception. However, the mathematical formalism developed there allows many physical interpretations. In the second paper, the analogies used provided a means of exploring the conceptual possibility that electric and magnetic actions are continuous actions (*physical* lines of force) taking place in a mechanical medium (*Maxwell's* conception of them). The field equations derived in that paper allow that the field could be produced by many underlying mechanisms; Maxwell later went so far as to say infinitely many. So, the particular mechanical analogies used there were not "brought forward as a mode of connexion existing in nature."[84] Let us now turn to that paper.

4.3 The "physical lines of force"

Maxwell's second paper, "On physical lines of force," is about Faraday's "physical lines of force", i.e., the conception that the lines of force actually exist and are essential to the description of electric and magnetic actions.[85] However, we must keep in mind that it only concerns Faraday's speculation in its *general* form. As we have discussed, Maxwell added the "physical hypothesis" that the lines are states of a mechanical medium, which he attributes to Faraday, but which we have seen was not Faraday's "favourite notion." I find this paper remarkable in that Maxwell, through the use of a mechanical analogy, which is not consistent and which probably would not even work locally, to say nothing of globally: (1) found the correct field equations for electrodynamic phenomena, consistent with all the known experimental results; (2) showed that if the equations were correct, there is a time delay in the transmission of electromagnetic actions; and (3) calculated the velocity of the transmission of such actions to be approximately the speed of light. If (2) and (3) were correct, they should lead to the discovery of new experimental results, which would prove incompatible with the action-at-a-distance conception.

In his introductory remarks, Maxwell expressed his dissatisfaction with the action-at-a-distance conception, attributing his dissatisfaction partially to the same vivid image of the iron filings which so influenced Faraday:

> . . . The beautiful illustration of the presence of magnetic force afforded by this experiment, naturally tends to make us think of the lines of force as something real, and as indicating something more than the resultant of two forces, whose seat of action is at a distance, and which do not exist there at all until a magnet is placed in part of the field. We are dissatisfied with the explanation founded on the hypothesis of attractive and repellent forces directed towards the magnetic poles, even though we may have satisfied ourselves that the phenomenon is in strict accordance with that hypothesis, and we cannot help thinking that in every place where we find these lines of force, some physical state or action must exist in sufficient energy to produce the actual phenomena.[86]

The object of this paper was:

> . . . to clear the way for speculation in this direction, by investigating the mechanical results of certain states of tension and motion in a medium, and comparing these with the observed phenomena of magnetism and electricity. By pointing out the mechanical consequences of such hypotheses, I hope to be of some use to those who consider the phenomena as due to the action of a medium, but are in doubt as to the relation of this hypothesis to the experimental laws already established, which have generally been expressed in the language of other hypotheses.[87]

Thus, the purpose of the analogy used in this paper was different than that of the first paper. There the purpose had been to allow a spatial representation of the intensity and direction of the lines of force. But now, the analogy was intended to allow an examination of *how* the lines, if they exist, could be produced, i.e., by what sorts of forces. The distinction between the purposes of the analogies is that between aiding in a 'kinematical' formulation, providing a spatial representation of the intensity and direction of the lines of force; and aiding in a dynamical formulation, providing an analysis of the generation of the field.

The paper is divided into three parts: Part I concerns magnetic phenomena, Part II is about electric currents and electromagnetic induction, and Part III deals with static electricity. In this paper, Maxwell skillfully exploited the potential of the method of "physical analogy" for providing insight into unknown phenomena. The analogy is elaborated and altered in each section to meet the requirements of the new phenomena under investigation. The fact that it is not entirely consistent is beside the point. The point is that it served as a guide to Maxwell's thinking in his formulation of the electromagnetic field equations.[88] One striking example is that it is very probable that Maxwell had not even intended to write Part III when he began his analysis and that he only began this most crucial part of the paper as far as the field concept is concerned after Part II was in press.[89] The analogy clearly took Maxwell further than he had intended initially and provided him with the basis for the desired link between electromagnetism and light.

Beginning with a consideration of magnetic actions, Maxwell maintained that a suitable analogy had to be able to account for four things: (1) a tension along the lines of force, (2) a lateral repulsion between the lines of force, (3) the occurrence of electric actions at right angles to magnetic actions, and (4) the rotation of the plane of polarized light, passed through a diamagnetic substance, by magnetic action. The latter two constraints had been established experimentally, while the former two were suggestions by Faraday as to how best to conceive the distribution of the lines of force in space. Maxwell, again, "poached among" Thomson's images and began with his analogy between electric and magnetic forces and "the displacements of the particles of an elastic solid in a state of stress."[90] Maxwell, it should be recalled, was himself an expert on elastic solids, having written one of his first papers on the subject.[91] The analysis given in that paper undoubtedly prepared him for the tasks of this one and influenced his results in important ways – something which has been largely ignored in the Maxwell literature.

Maxwell showed that a mechanical analogy, consistent with the four constraints, is that of a fluid medium, composed of elastic cells, or "vortices," under a state of stress. The attempt to ascertain possible mechanical causes of such a stress led to his formulation of the field equations. Generally, as Maxwell said, stress is conceived of as "action and reaction between the consecutive parts of a body, and consists in general of pressures or tensions different in different directions at the same point of the medium."[92] The stresses involved in magnetism must consist of (1) a tension in the direction of the lines of force, because in both attraction and repulsion the object is drawn in the direction of the resultant of the lines of force and (2) a pressure greater in the equatorial than in the axial direction, i.e., a hydrostatic pressure, because of the lateral repulsion of the lines of force. The excess of pressure in the equatorial direction can be explained as the result of the centrifugal force acting on the vortices, which are in the plane perpendicular to the lines of force, rotating in a clockwise direction with axes parallel to the lines. (This is consistent with Faraday's demonstration that magnetic actions involve rotation and with Thomson's representation of magnetic forces as rotational strains in an elastic solid.) The lines of force can thus be construed as indicating the direction of least pressure. Also, the dipolar nature of magnetism can be accounted for since each extremity of the axis of a vortex is distinguished by its direction of rotation.

Given the above considerations, Maxwell derived an expression for the resultant force on an element of the medium due to its internal stress. The components of this expression form what we today call the 'electromagnetic stress tensor'.[93] Given the stress tensor, it is possible to calculate the resultant force on any portion of the medium. Maxwell then found that the intensity of the magnetic force could be conceived of as proportional to the velocity of the circumference of the vortices and that the specific magnetic inductive capacity, i.e., 'magnetic permeability', is proportional to the density of the vortices. Thus, the 'magnetic quantities' could be represented.

Using the analogy, Maxwell was able to give representations for magnetic induction, paramagnetism, and diamagnetism. He also showed that in the limit where the magnetic permeability is uniform and where there are no electric currents, his analysis gives the same result as action at a distance. Though his results thus far did not support the field conception over the other, what he had achieved was important. With the use of a mechanical analogy, he had shown that it is possible to give mathematical formulation to the conception of magnetic actions as continuous processes. He next used the analogy to give an account of the relationship between currents and magnetism.

The problem of how to construe electric currents required a modification of the analogy. The solution was easy: Contiguous parts of the vortices must be going in opposite directions since they all rotate in the same direction; thus, the vortices would stop. Mechanical consistency requires the introduction of 'idle wheels' surrounding the vortices. Maxwell conceived of these 'idle wheels' as small spherical particles, surrounding the vortices and revolving in the direction opposite to the motion of the vortices, without slipping or touching each other. Their translational motion could represent an electric current. Given the analogy, a current would produce magnetic effects in the following way. When an electric force acts on the particles, it pushes them and starts them rolling. The tangential pressure between them and the vortices sets the vortices in motion, or alters their motion. In this way, also, changes in the motion of the vortices would produce motion of the particles, i.e., currents in a conductor. Maxwell showed that these 'relationships' between currents and magnetism could be formulated mathematically as the relationship between the motion of the particles and the circumferential velocity of the vortices. The complete analogy was represented by Maxwell in Figure 2.[94]

Maxwell next used the analogy to derive the laws of electromagnetic induction. The derivation was in two parts: First, he found the electromotive force on a stationary conductor, produced by a changing magnetic field; second, he found the electromotive force on a conductor moving through a magnetic field. Given the analogy, one can see quite clearly that a changing magnetic field would produce a current. Maxwell's derivation

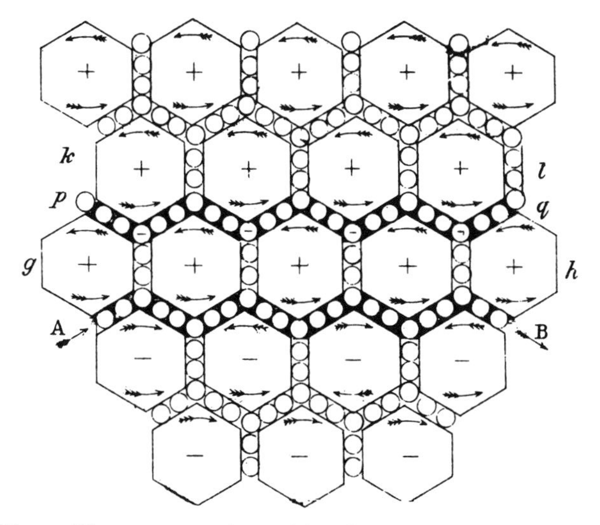

Figure 2. Maxwell's representation of his "physical analogy" (from Maxwell (1861–2)).

of the mathematical relationship came from a consideration of the energy of the medium due to the motion of the vortices. The energy is proportional to the density of the medium and the square of the circumferential velocity of the vortices. Change in the energy is related to change in magnetic force. This relationship allowed the derivation of the equation connecting the induced electromotive force and change in magnetic force. Maxwell, at this point, reestablished the relationship between the magnetic field and the "electrotonic function," which he had derived in the previous paper. The analogy now provided him with a solution to the major unresolved problem of that paper: how to represent Faraday's "electro-tonic state." The "electro-tonic state" could be represented as the "reduced momentum" of the vortices; that is, the "electro-tonic state" is represented as the state of stress which would be produced if the entire mechanism suddenly came to a halt.

Turning now to the second part, we can see that motion of a conductor through the magnetic field would alter the motion of the vortices, giving rise to an electromotive force which would set the particles of the conductor in motion, producing a current. The motion of the conductor stretches the vortices, causing them to speed up. As the conductor moves along, the vortices behind contract and slow down. The difference in vortex velocity between those in front of and those behind the particles in the conductor gives rise to a force which 'pushes' the particles. The expression which Maxwell derived for the total force on a conductor moving through the medium is composed of three factors: (1) the motion of the conductor, (2) the change in the "electro-tonic state," and (3) the electric tension due to static charge.[95] Factor (3) enters into the equation for purely formal reasons. The interpretation of it as "electric tension" is merely asserted by Maxwell, and it seems to have no justification in terms of analogy.

Maxwell related his formulation of electromagnetic induction to that of Faraday, which, it will be recalled, involved a relationship between the strength of the induced current and the number of lines of force crossed per unit time.[96] As we saw in our discussion of Faraday, the quantitative relationship given in terms of the number of lines cut is correct only in rare cases. Maxwell's mathematical formulation goes beyond the lines of force 'measure' he, himself, attributed to it here and is not subject to that objection.[97]

Maxwell's results thus far show that a field formulation of electromagnetic phenomena is possible and that this formulation is consistent with known experiments. Impressive as these results are, they still do not provide a crucial counter to the action-at-a-distance conception. This was to come in Part III. The analogy developed in the previous sections suggested to Maxwell a way to represent and to give mathematical formulation to electrostatic induction. The introduction of what Maxwell called the "displacement current" into the equation led to the conclusion that, if the equation were correct, the transmission of electromagnetic actions takes time.

Maxwell's application of his mechanical analogy to electrostatics and charge again illustrates the heuristic value of his method. In his attempt to extend the analogy beyond his original intentions, i.e, beyond the attempt to represent magnetism, currents, and electromagnetic induction, he was led to hoped for – yet what he himself called "surprising" – results. In this part of the analogy, Maxwell made several, fortuitous, errors: in sign and in the equation for transmission of transverse effects in the medium. Additionally, he did not follow out some of the seemingly obvious conclusions that could be drawn from the use of the analogy in this paper, but waited until his next, more formal, paper to formulate them. Because of these facts, some writers have tended to play down the importance of the analogy in Maxwell's reasoning; with one, Duhem, going so far as to accuse Maxwell of "falsifying" one of his results and of constructing the model *after* he had derived the equations.[98] I will argue that the major 'errors' follow directly from the analogy and from his previous work on elasticity, with no 'fudging' necessary, and that the conclusions most likely were omitted because of Maxwell's conception of the method of physical analogy.

In Part III, Maxwell altered the analogy by making the entire medium elastic.[99] He did so because "a conducting body may be compared to a porous membrane which opposes more or less resistance to the passage of a fluid, while a dielectric is like an elastic membrane which may be impervious to the fluid, but transmits the pressure of the fluid on one side to that on the other."[100] So, in a dielectric, an actual current is not produced, but a polarization, creating a 'tension', does occur. Maxwell considered charge, itself, to be a manifestation of an excess of polarization in the medium and, thus, to be spread out in space. Given the analogy, polarization of a dielectric could be represented by both the cells and the particles remaining stationary but being capable of elastic distortion or 'displacement'. Maxwell likened the effect of the displacement to a current – a "displacement current":

> This displacement does not amount to a current, because when it has attained a certain value it remains constant, but it is the commencement of a current and its variations constitute currents in the positive or negative direction, according as the displacement is increasing or diminished.[101]

The notion of a "displacement current" led to considerable confusion among the readers of his paper. What was meant is that the existence of a displacement in the particles is not a current, nor does it have the effects of one, but any *change* in the displacement does "amount to a current." That is, the electrical particles, when polarized, are distorted in place. Any change in displacement produces a tangential force on the spherical cells, which is then communicated to the particles and cells throughout the medium, producing 'current-like' motion.

Maxwell derived the equation representing the relationship between the induced electromotive force and the displacement by analogy with displacement in elastic solids. This equation led to some confusion and still creates problems in trying to understand what he had intended.[102] Maxwell's sign conventions are not what is now customary. He appears to take "displacement" to be in the *opposite* direction from the field intensity and not, as is now usual and as he later held, in the same direction as the field intensity. It strikes me that given the analogy Maxwell's equation is correct: If the electromotive force is represented by the 'restoring force' in an elastic solid, then the sign is correct, since that force and the displacement would have opposite orientation.[103] Without the analogy, there is no justification for a sign difference. Through a later mistake in sign, he did arrive at the customary equation for the 'total current', i.e., the ordinary current *plus* the displacement current. Unlike the currents considered thus far, the total current is now 'open', i.e., not circuital. There are problems of interpretation here as well, since Maxwell was careless with his use of subscripts, but they can be resolved with careful analysis. What is surprising is that Maxwell did not discuss the important result that, given the equation for the total current, the circulation of the magnetic field is now related to the total current and not only to the conduction current. This new formulation implies that *static* electricity produces magnetic effects.[104] Maxwell did specify this in the next paper, and it is hard to believe that he did not notice the possibility here. In the absence of any mention of this new possibility, it is difficult to say anything definite about why he did not. Two possibilities suggest themselves. First, given the analogy, it is unclear whether the distortion of the vortices should be considered as a slight rotation, both when it is initiated and when it ceases. Second, and this is also pertinent for the second crucial conclusion he failed to draw explicitly in this paper, if the physical analogy is to be regarded in the way Maxwell advocated, then it can, at most, *suggest* physical hypotheses but cannot be the basis for advocating any. There was no empirical evidence that electrostatic actions produce magnetic effects, and there was also, at this point in Maxwell's thinking, no obvious way to derive the relationship independently of the analogy. Let us now discuss the second omitted conclusion.

Maxwell interpreted the constant factor in the equation relating the electromotive force and the displacement as the velocity of transverse vibrations in the medium. An ordinary elastic medium will transmit both longitudinal and transverse waves, which will in general have different velocities. Maxwell confined his attention to transverse waves, perhaps since light was known to consist in transverse vibrations of its medium and he wished to compare the two velocities. He calculated the value of the constant using the electromagnetic measurements of Weber and Kohlrausch, which on his analysis determine the speed of transverse vibrations in the electromagnetic medium. The result proved to be somewhat of a surprise to Maxwell: The velocity agreed very closely with the value of the speed of light. Again, Maxwell made a fortuitous error in this calculation: He used the wrong equation for the velocity. Given his analogy, the velocity is equal to the square root of the coefficient of rigidity of the medium divided by *twice* the density, rather than simply the density, which Maxwell used. The mistake can be traced to Maxwell's earlier paper on elasticity, where he gave the same formulation for the velocity.[105] It is most likely that Maxwell simply did not reformulate the equation for the model and took it from his previous work, rather than deliberately falsifying the result as claimed by Duhem.[106] If this interpretation is correct, Maxwell was extremely lucky. However, he was soon to have more concrete results anyway, independent of the analogy, in his attempt to determine international electrical standards for the British Association.[107] There he determined that the ratio of the forces between electric charges and of the forces between magnetic poles is of a magnitude very nearly that of the velocity of light.

I said above that the result was "somewhat of a surprise" because, although it had not been the goal of this paper, Maxwell had hoped to produce a unified theory of electromagnetism and light. However, he *was* surprised that this calculation produced the desired result immediately. As he said in his letter to Thomson, "I made out the equations in the country before I had any suspicion of the nearness between the two values of the velocity of propagation of magnetic effects and of light, so that I think I have reason to believe that the magnetic and luminiferous media are identical . . . ."[108] In the paper he concluded, quite excitedly, that "we can scarcely avoid the inference that *light consists in the transverse vibrations of the same medium which is the cause of electric and magnetic phenomena.*"[109] Yet he did "avoid the inference" until his next paper, in which he derived the wave equation for electromagnetic phenomena and recognized that light is itself an electromagnetic phenomenon. Again, I think we have to see his reticence as an attempt to apply the method of physical analogy strictly. His value for the velocity is related directly to the physical analogy, but the analogy is not a *physical hypothesis.* That is, Maxwell was not claiming that the electromagnetic medium is an elastic medium, filled with vortices and idle wheels. He simply was maintaining that the stresses and strains which produce electromagnetic phenomena could be represented by such a medium. What he had to say concerning the introduction of the electrical particles into the analogy, he also meant for the whole of the analogy:

> *I do not bring it forward as a mode of connexion existing in nature, or even as that which I would willingly assent to as an electrical hypothesis* [italics mine]. It is, however, a mode of connexion which is mechanically conceivable, and easily investigated, and it serves to bring out the actual mechanical connexions between the known electromagnetic phenomena; so that I venture to say that *any one who understands the provisional and temporary character of this hypothesis, will find himself rather helped than hindered by it in his search after the true interpretation of the phenomena* [italics mine].[110]

So, given that the analogy was not conceived of as a "mode of connexion existing in nature," Maxwell could not very well claim that his calculation of the velocity of transverse vibrations in such a medium actually gives the correct (i.e, real) velocity for electromagnetic actions. He needed a way to determine the velocity without the analogy, and this he found in his next major paper on electromagnetism, "A dynamical theory of the electromagnetic field."[111] In that paper, Maxwell significantly changed his method, as will be discussed shortly. He 'scrapped' the mechanical analogy developed in this paper and, using known relations and experimental results: (1) derived the electromagnetic field equations, (2) deduced the electromagnetic theory of light, and (3) demonstrated that highly complex interactions should take place between electric and magnetic actions in space. Before turning to that paper, though, let me add a few concluding remarks about the "physical lines" paper.

What Maxwell presented in this paper is a field formulation of electromagnetic actions, which indicated that their transmission time has finite value, based on Faraday's concept of physical lines of force. Here, also, there is a divergence from Faraday's conception in that the field is not propagated along its lines of force, but in a direction orthogonal to them, so that the lines are neither the 'vehicles' nor the 'paths' of transmission of the actions. However, Faraday's *general* speculation is vindicated by this analysis: Mathematical formulation of the physical-lines conception demonstrated that, if the equations were correct, the suspected time delay did exist and should be detectable experimentally. Thus, the field concept enters essentially into the description of electric and magnetic actions. The results of this paper, of course, did not amount to a refutation of the action-at-a-distance conception – that had to await new experimental results. What Maxwell did achieve was: (1) He gave strong reasons for doubting that conception, i.e., for regarding it as likely that the actions involve a time delay and (2) he provided a viable alternative to it, which he, himself, considered preferable.

Maxwell had achieved his results with the aid of a mechanical analogy which he did not put forward as a "physical hypothesis." He later was to say, with regard to this analogy, that: "The problem determining the mechanism required to establish a given species of connexion between the motion of the parts of a system always admits of an infinite number of solutions."[112] Still he maintained that certain conditions such solutions must meet were established in this paper:

(1) Magnetic force is the effect of centrifugal force of the vortices.
(2) Electromagnetic induction of currents is the effect of the forces called into play when the velocity of the vortices is changing.
(3) Electromotive force arises from stress in the connecting mechanism.
(4) Electric displacement arises from the elastic yielding of the connecting mechanism.[113]

Given these conditions, which are mentioned near the end of his life's work, and taking into account everything that has been said thus far about its beginning, we can see that he did make one crucial physical hypothesis, which he always adhered to: It is mechanical (i.e., Newtonian) forces in an underlying medium which produce electric and magnetic phenomena. That is, the electromagnetic field is a state of a Newtonian aether. Given this hypothesis, his use of a detailed mechanical analogy in this paper and of general dynamical (Lagrangian) considerations in the next was entirely appropriate. The analogy was of great heuristic value in conceptualizing and in formulating mathematically the relationships between the various stresses and strains thought to be at work in electricity and magnetism. The use of generalized dynamics, which will be discussed more fully in the following section, enabled him to derive the field equations and to deduce the electromagnetic theory of light in a way that was both more straightforward and more acceptable to the physics community, while at the same time leaving open the nature of the underlying mechanism. Actually, his use of generalized dynamics masked the fact that the electromagnetic field equations are those of a system to which Newton's laws do not apply!

4.4 'Mechanical' to 'dynamical'

Before discussing the major results of Maxwell's third paper, we need to see its importance as marking a transition in the *process* of the development of the concept of field. Up to this point, the use of analogy with 'known' phenomena has been crucial to the development of the field conception of electric and magnetic actions. With Faraday, we have the use of rather vague analogies which provided him with a way to begin to conceptualize how such actions could be transmitted continuously through space, along the lines of force. With Maxwell, we have the very detailed and mathematically sophisticated "physical analogies" which acted as heuristic guides allowing him to explore the new conceptual possibility and enabling him to derive the correct equations for the electromagnetic field. Beginning with this third paper there are no more detailed mechanical analogies concerning the underlying structure which could produce the electromagnetic field. From this point on, Maxwell used only general analogies the purpose of which was primarily "illustrative."[114] As he said in the paper, he now wished "merely to direct the mind of the reader to mechanical phenomena which will assist him in understanding the electrical ones."[115] Once the correct laws had been obtained, there was no further need to draw upon results in other domains of physics. This, of course, was even more true for the followers of Maxwell. Electromagnetic field theory had become a scientific domain in its own right, and there was little need to borrow insights from other fields. What was needed now was a clarification, elaboration, and extension of the new mathematical formalism. Maxwell's third paper, in which he produced, among other things, the electromagnetic theory of light, marked the beginning. The most important thing to note for our purposes is that during this 'elaborational' period of development, a significant change in the concept of field did occur (in the work of Lorentz), which was largely the result of the attempt to solve mathematical problems posed by both mathematical and physical constraints on the field and its interaction with matter. This will be discussed in the next section.

The transition within Maxwell's own work, as marked by the contrast between his second and his third paper, involved the change from conceiving of the electromagnetic field as a 'mechanical' system to conceiving of it as a 'dynamical' system. 'Mechanical' here means a Newtonian system, and 'dynamical' means a general dynamical (Lagrangian) system. This transition was crucial to the development of the conception of electromagnetic actions: partly because it provided a means of deriving the field equations independent of the mechanical analogy in the previous paper, and partly because the Lagrangian formalism provides no information about the underlying situation. The latter allowed Maxwell and others to proceed without being hampered by having to consider the nature of the 'causes' of electromagnetic phenomena. Maxwell argued that, following Newton's lead with gravitation, it is possible, and sometimes necessary for the progress of science, to formulate the laws of phenomena without specifying the details of the underlying mechanism thought to produce such phenomena. He attributed to Newton the methodological principle that "we should investigate the forces with which bodies act on eachother in the first place, before attempting to explain *how* that force is transmitted."[116]

Generalized dynamics provided Maxwell with the means for following through with this principle. When we lack detailed information about the phenomena under investigation:

> The question is not – what phenomena will result from the hypothesis that the system is of a certain specified kind? but – what is the most general specification of a material system consistent with the condition that the motions of those parts of the system which we can observe are what we find them to be?[117]

The knowledge of the system required in order to apply generalized dynamics is minimal: velocity, momentum, potential energy, and kinetic energy. With this information, it is possible to derive the equations of motion for any "connected system." Maxwell discussed the common features of such systems in many places.[118] I will apply what he had to say to the case at hand. First, the system under consideration is the *electromagnetic field*, "because it has to do with the space in the neighbourhood of the electric or magnetic bodies."[119] Second, it is a *dynamical system* in that "it is assumed that in space there is matter in motion by which the observed electromagnetic phenomena are produced."[120] Finally, it is a *connected system* since "the motion [is] communicated from one point of the system to another by forces."[121]

Maxwell made a useful comparison between the use of generalized dynamics in the case of a connected system and the situation in a belfry in which the bellringers cannot see the mechanism which rings the bells. Its clarity makes it worth quoting in its entirety:

> In an ordinary belfry, each bell has one rope which comes down through a hole in the floor to the bellringer's room. But suppose that each rope, instead of acting on one bell, contributes to the motion of many pieces of the machinery, and that the motion of each piece is determined not by the motion of one rope alone, but by that of several, and suppose, further, that all this machinery is silent and utterly unknown to the men at the ropes, who can only see as far as the holes in the floor above them.
>
> Supposing all of this, what is the scientific duty of the men below? They have full command of the ropes, but of nothing else. They can give each rope any position and any velocity, and they can estimate its

momentum by stopping all the ropes at once, and feeling what sort of tug each rope gives. If they take the trouble to ascertain how much work they have to do in order to drag the ropes down to a given set of positions, and to express this in terms of these positions, they have found the potential energy of the system in terms of the known co-ordinates. If they then find the tug on any one rope arising from a velocity equal to unity communicated to itself or to any other rope, they can express the kinetic energy in terms of the co-ordinates and velocities.

These data are sufficient to determine the motion of every one of the ropes when it and all the others are acted on by any given forces [italics mine]. This is all that the men at the ropes can ever hope to know.[122]

The comparison with the electromagnetic situation is obvious. Once we determine the generalized coordinates of the system, it is possible to determine the configuration of the field at any moment. However, the underlying mechanism which actually produces the field remains unknown.

In concluding this discussion of generalized dynamics, let me just add the following consideration. I have characterized the general dynamical formulation of electromagnetism as the transition from a 'mechanical' representation of the aether to that of a 'dynamical'. Given our present understanding of 'dynamical', this leaves open the possibility of the aether being a non-Newtonian system, while still being a dynamical system. This is so because we use 'dynamical' in a wider sense than in the period under discussion. For us, many dynamical systems can be formulated in general dynamical terms, e.g., relativistic mechanics. However, historically, analyses leading to the formulation of generalized dynamics began with Newton's equations of motion: Such a formulation was seen as a more general way of characterizing a mechanical, i.e., *Newtonian*, system; in particular, systems about which there is little information. So, 'dynamical' and 'mechanical' were taken as co-extensive. In Maxwell's view it was inconceivable for the aether to be a non-Newtonian substance. Also, even with the dynamical formulation of electromagnetism, just what kind of a mechanical system constitutes the aether remained a major concern for him.[123] What he continued to maintain is that his laws are valid *regardless* of the nature of the mechanism which actually produces the phenomena. Indeed, the laws act as a constraint on any acceptable mechanism. But, for Maxwell, knowledge of that mechanism was an essential part of a complete dynamical theory of electromagnetism, and thus he regarded his own theory as incomplete.

Coming at last to the third paper, clearly the major problem facing Maxwell was the derivation of the electromagnetic theory of light. He was to achieve this in Part VI, after he had rederived the field equations. He began the paper by assuming the existence of the luminiferous aether:

> . . . an aethereal medium filling space and permeating bodies, capable of being set in motion and of transmitting that motion from one part to another, and of communicating that motion to gross matter so as to heat it and affect it in various ways.
>
> . . . the parts of this medium must be so connected that the motion of one part depends in some way on the motion of the rest; and at the same time these connexions must be capable of a certain kind of elastic yielding, since the communication of motion is not instantaneous, but occupies time.[124]

Thus, the luminiferous aether is a 'connected system'. It is an elastic medium which is capable of receiving, storing, and transmitting energy. He then argued that it is reasonable to assume the existence of an electromagnetic aether, similar to the luminiferous, taking as support the observation that the rotation of the plane of polarized light is affected by magnetic actions and his "theory" that the medium is capable of "a kind of elastic yielding." By assuming the elasticity of the electromagnetic medium, Maxwell built the time delay of electromagnetic actions into his conception. The problem now was to find a way to calculate the velocity of such actions and then compare that velocity with the velocity of light. He concluded that the electromagnetic system is one to which generalized dynamical reasoning applies because it is a 'connected system', with energy, potential and kinetic, existing *in* the medium.[125] This is in marked contrast to the action-at-a-distance conception:

> In speaking of the Energy of the field, however, I wish to be understood literally. All energy is the same as mechanical energy, whether it exists in the form of motion or in that of elasticity, or in any other form. The energy in electromagnetic phenomena is mechanical energy. The only question is, where does it reside? On the old theories it resides in the electrified bodies, conducting circuits, and magnets, in the form of an

> unknown quantity called potential energy, or the power of producing certain effects at a distance. *On our theory it resides in the electromagnetic field, in the space surrounding the electrified and magnetic bodies, as well as in those bodies themselves* [italics mine], and in two different forms, which may be described without hypothesis as magnetic polarization and electric polarization or, according to a very probable hypothesis, as the motion and strain of one and the same medium.[126]

So, the field interacts with ordinary matter by transmitting energy to and receiving it from it. Thus, a crucial dynamical property, energy, is attributed to the field, making the field concept essential to the description of electric and magnetic actions. One of the major problems with Maxwell's theory, though, is that there is no account of how the field and ordinary matter interact. This is a problem others were to attempt to resolve after Maxwell.

Given both the foregoing assumptions and certain results of his previous papers, Maxwell derived the general equations of the electromagnetic field (eight in his version) – with the circulation of the magnetic field now related to the *total* current – and the expression for the total energy of the field.[127] He then applied them to specific problems. The details of the derivations are not important for us, but two of the applications are. First, he determined the mechanical force acting on an electrified body, which provided a means of determining the coefficient of "electric elasticity": a velocity of transverse propagation, and equal to that of light. Second, he showed that the "properties of that which constitutes the electromagnetic field, deduced from electromagnetic phenomena alone, are sufficient to explain the propagation of light through the same substance."[128] He derived the wave equation for electromagnetic phenomena from the field equations in a purely formal way, by coupling the equation for the circulation of the electric field with that for the circulation of the magnetic field.[129] He then showed that the velocity of such waves in a vacuum is approximately that of light and that, like light, only transverse waves are possible. The latter was a significant achievement as well, since the elastic solid theory of light had encountered considerable problems in attempting to explain away the longitudinal waves such a medium should allow, but which had never been observed. In sum, Maxwell had established that the electromagnetic theory of light is independent of any specific assumptions concerning the nature of the medium of transmission. This time, Maxwell did not avoid the inference: "The agreement of the results seems to shew that light and magnetism are affections of the same substance, and that light is an electromagnetic disturbance propagated through the field according to the electromagnetic laws."[130] He had also demonstrated, with the derivation of the wave equation, that highly complex interactions should take place between electric and magnetic fields in space, in the absence of ordinary matter.

Later, in the *Treatise*, Maxwell formulated electromagnetism more fully in general dynamical terms.[131] He provided a very 'modern' analysis which far outstripped the use of generalized dynamics by his contemporaries. Since no change occurred in the concept of field in this analysis, there is no need to discuss it here. However, it should be noted that his analysis was, and is, very difficult to understand, and, unfortunately, it was through the *Treatise* that continental scientists became acquainted with Maxwell's work.

4.5 Summary: Maxwell's concept of field

What, by way of summary, can be said about the concept of field as it emerges from Maxwell's work? First, it has now become indispensable for the description of electric and magnetic actions.[132] There are two separate, but interpenetrating and interacting fields: the electric field and the magnetic field. A complete inventory of bodies and their motions is not sufficient to give a complete characterization of the nature of the interactions involved. The concept of field enters essentially into the laws of electrodynamics because of the time delay in transmission. That is, physical processes must be taking place in the intervening space. As a consequence of the time delay, the dynamical properties of energy and momentum, formerly applied only to matter (ponderable), must be attributed to the field in order that these quantities be conserved. However, we must remember that in Maxwell's conception the field does not have ontological status co-equal with matter; it is not a substance, but a state of a substance, the aether. As I characterized it in the beginning of this section, it is a *Newtonian aether-field.* Maxwell saw the construction of a *mechanical* representation of its action as one of the central problems of electromagnetic theory.

In conclusion, we have seen that Maxwell took Faraday's lines-of-force conception of electric and magnetic actions, gave it mathematical formulation, and, in so doing, provided a powerful alternative to the action-at-a-distance conception. What Maxwell did not realize is that the mechanical representation he wanted for the "underlying processes" was not possible because in fact he had formulated the laws of a *non*-Newtonian dynamical system. It is also important to remember that in giving mathematical formulation to each aspect of Faraday's conception, the "representative" and the "physical," Maxwell at the same time altered both. Regarding the former, Maxwell changed the relationship formulated by Faraday between the intensity of the force and the number of lines cut, by replacing 'number of' with a continuous measure. As to the latter, with Maxwell's electromagnetic field, the lines of force are neither the 'vehicles' nor the 'paths' of transmission of the actions. These changes do not, however, mark a radical 'break' between the conceptions of Faraday and Maxwell. Maxwell's conception and Faraday's conception agree in their general form: The lines of force represent some state of, or are due to some process in, the intervening space and that process may be of fundamental importance for the theory of electric and magnetic actions. What Einstein had to say about the relationship between the work of Faraday and that of Maxwell provides an accurate characterization:

> . . . the pair Faraday-Maxwell has a most inner similarity with the pair Galileo-Newton – the former of each pair grasping the relations intuitively, and the second one formulating those relations exactly and applying them quantitiatively.[133]

Hendrik Antoon Lorentz (1853–1928)

CHAPTER 5

Lorentz' 'non-Newtonian aether-field'

5.1 An enigmatic scientist

In moving straight from Maxwell to Lorentz and then on to Einstein, we are obviously passing over a long list of aether theorists of late-19th and early-20th century physics. I find this period intriguing to study in that some of the best mathematical and scientific minds that have ever lived devoted their entire lives to the construction of theories about an entity which, we would now say, did not exist. They were engaged in what Illy has aptly called "the construction of a substance from its attributes."[134] Although there are some good analyses of portions of this period, it is not possible to assemble a complete fine-structure analysis from the existing literature. However, providing such an analysis would go far beyond the scope and purpose of this book. Although there were subtle differences in the meanings associated with the different views concerning the properties of the aether, as far as I have been able to discern, the aether was considered a Newtonian substance by all, with one exception: H. A. Lorentz. For Lorentz, the field became a state of a non-Newtonian aether. I say 'became' because he did not set out to produce a non-Newtonian conception; it was simply a by-product of the separation he made between the aether and 'ordinary matter'.

Discussion of Lorentz' contribution presents us with difficulties. First, the details of how he arrived at his conception of the field as a state of a non-Newtonian substance are difficult to put in an easily comprehensible form for the reader lacking sufficient technical background.[135] Lorentz did not rely on visual images in his reasoning – at least as far as he indicates in his writings. Also, the use of "physical analogies" was no longer necessary since electromagnetic field theory had become a scientific domain in its own right. We saw this transition take place at the point where Maxwell's analysis went from the heavy use of "physical analogies" in his second paper to the use of very general analogies and generalized dynamics in the third paper and the *Treatise*. Lorentz discussed his own reluctance to make use of visualization in physics in his inaugural address at Leiden:

> But one can also have too much of a good thing and, thus, . . . by visualizing too much one can overshoot the mark, and place so much emphasis on what should serve as a picture, that it is taken too much for the thing itself . . . .
>
> Now one must especially guard oneself against such an excess of visualization when one is concerned with *forces* in physics.[136]

He went on to say that "the word 'force' is just a name for some quantities which occur in our mathematical formulae."[137] In his work, we are not presented with "pictures" of how electric and magnetic forces are transmitted.

Second, Lorentz presented his work in what could be called 'modern journal form'; that is, without more details of how he had arrived at a certain conception or of why he had taken a particular approach than were absolutely necessary to following the mathematical argument. He was not given to making comments about the way he had arrived at his ideas – not even in his unpublished scientific notebooks! As Casimir has noted, "Lorentz did not like to speak about a problem before he had arrived at a solution."[138]

Lorentz also did not provide any significant elaboration of, or reflections on, either his metaphysical presuppositions or his conception of the nature of the scientific enterprise. It is not clear whether he simply never had such reflections or was reluctant to express them. De Haas, his son-in-law and colleague, said of him:

> Lorentz did not like to discuss things not well known to him, or matters to which he had neither given, nor tried to give his deeper thoughts. He never touched philosophy.[139]

However, the late correspondence between Lorentz and Einstein, which is obviously a continuation of discussions between the two, provides evidence that Einstein, at least, was able to elicit philosophical reflections from him. It is interesting to note that it is Einstein who appears to have appreciated best the significance of Lorentz' contributions and their radical nature:

> The physicist of the present generation regards the point of view achieved by Lorentz as the only possible one; at the time, however, it was a surprising and audacious step, without which the later development would not have been possible.[140]

We will examine the nature of the "surprising and audacious" steps Lorentz made as they relate to the formulation of his electromagnetic field concept. The importance of what has been said thus far for our examination of those steps is to see that, for Lorentz, philosophical reflections are not part of the scientific enterprise as such. Thus, he could easily dismiss objections to the non-Newtonian aspects of his conception by such remarks as "But why must we in truth, worry ourselves about it?", while at the same time he could fail to make the final step to Einstein's conception of field because he never reflected on his own adherence to the Newtonian conceptions of space and time.[141]

One final point needs to be made here in order not to misinterpret some of the discussion which follows. Lorentz' attitude towards doing science could be called 'instrumental', but it would be a mistake to call him an 'instrumentalist' in the standard philosophical use of the term. His attitude is 'instrumental' in that he held that we do science by making and holding hypotheses that work, without worrying about whether they are 'true', until they no longer work or until a better, more useful one comes along. He was even willing to make use of what he, himself, considered "startling" hypotheses that seemed to solve problems, such as his hypothesized contraction of matter. The important thing was to realize the provisional nature of such hypotheses. What he said in a lecture on the aether illustrated his attitude:

> But, allow me to remind you that without hypotheses scientific explanation is unthinkable. Newton's '*hypotheses non fingo*' must certainly be taken *cum grano salis* and though Ampère thought he had deduced the laws of electrodynamics from experience alone, he forgot a supposition

> upon which his whole edifice rested. We will always make hypotheses; we have only to watch out that we do not take too much pleasure in this use of our imagination.[142]

However, he cannot be called an 'instrumentalist' in the sense of not making "physical hypotheses" (in Maxwell's terminology). Lorentz certainly did bring forward some of his hypotheses "as a mode of connexion existing in nature," in particular his hypothesis of electromagnetic phenomena as involving the interaction of a stationary aether with movable charged particles. It is from the hypothesis of the stationary aether that the non-Newtonian consequences followed. Let us now look at the development of his field conception.

5.2 Rapprochement

Lorentz, quite early in his work, created a *rapprochement* between the action-at-a-distance and the continuous-action conceptions of electromagnetic actions. He achieved this in attempting to resolve the central problems facing electrodynamics: (1) The overall motion or lack of motion of the aether, (2) the nature of the aether–matter interaction, and (3) the nature of charge. However, as we will see, although this *rapprochement* was a great achievement, its usefulness was to be short lived because it incorporated the fundamental conflict between matter obeying Newton's laws and fields obeying Maxwell's laws without actually resolving it. Lorentz did not create a unified theory of mechanics and electrodynamics, as Einstein was to do. Rather, his position was that electromagnetic actions need not fit into the Newtonian framework. He arrived at this position as a consequence of his hypothesis that the aether is stationary. First, and most importantly, if the aether is immobile, i.e., if every part of the aether is permanently at rest, then it is impossible to give any underlying *mechanical* representation for electromagnetic forces. Thus, the Maxwellian claim that we could simply proceed in the manner of Newton, trusting that at some point a mechanical explanation would be given, was no longer tenable.[143] Second, if a mechanical explanation could not be given in principle, then there was no reason to assume *a priori* that electromagnetic actions are subject to Newton's laws.

Lorentz was in a good position to create a *rapprochement* because he started his work with a modified action-at-a-distance conception of electromagnetic actions, i.e., that of Helmholtz, and only later accepted the continuous-action conception. Helmholtz had found Maxwell's electromagnetic theory of light attractive, but difficult to understand in the manner Maxwell had presented in his *Treatise.* He, thus, provided an action-at-a-distance–aether reformulation of Maxwell's theory, which was more appealing and comprehensible to continental physicists. Helmholtz' conception, broadly characterized, is that electromagnetic actions take place through the polarization of aether particles, which act on one another at a distance.

In his discussion of Maxwell's theory, Helmholtz had stated that one of the greatest problems of the electromagnetic theory of light concerned the reflection and refraction of light, which Maxwell, himself, had not considered.[144] Lorentz took this question as the starting point for his dissertation.[145] Thus, optical phenomena were at the center of Lorentz' concerns from the outset and, we should keep in mind, were to remain primary in his thinking. Also, although he had read Maxwell, he preferred to adopt Helmholtz' action-at-a-distance interpretation because it had "the advantage of founding the theory on the most direct interpretation of the facts."[146] The major conclusion of Lorentz' dissertation was the superiority of the electromagnetic theory of light over the elastic solid theory. He demonstrated conclusively that it would not be possible to suppress longitudinal waves in such a medium. He then argued that since such light waves had never been observed, Maxwell's theory, which predicts only transverse waves, is superior.

In the third chapter of this work we find a rudimentary form of Lorentz' later separation between aether and ordinary matter. In dealing with the question of the nature of electromagnetic phenomena in gases, he considered gas molecules to be tiny particles embedded in the aether. He argued that electromagnetic phenomena in gases are due mainly to the action of the aether and very little to that of the gas molecules. For support he used the observation that the velocity of light in, and the inductive capacitances of, gases and the vacuum are nearly the same. He then extrapolated his conclusion to the cases of liquids and solids. It was not until his next paper that he was to make an explicit and clear distinction between the role of the aether in optical and electromagnetic phenomena in bodies and the role of the particles which constitute bodies.[147]

A significant problem with Maxwell's theory is that there is not a clear separation between the aether and other dielectric substances. The aether is a special instance of a dielectric body. Thus, there is a problem with how to conceive of electromagnetic fields within matter. The aether, on Maxwell's account, has different properties in a vacuum and within bodies. Another significant problem is how to conceive of charge. As Lorentz noted, "Poincaré mentions a physicist who declared that he understood the whole of Maxwell's theory, but that he still had not grasped what an electrified sphere was"![148] Lorentz made a definite separation between aether and matter, which led to a clear understanding of 'field' and 'charge'. In his 1878 paper, the aether is the only true dielectric. It penetrates matter and is present between the molecules of rigid bodies and fluids, as well as gases. He maintained that only with this assumption could we account for the propagation of light and other optical phenomena in matter. This all-pervasive aether has the same properties in a vacuum and within matter and is responsible for *all* dielectric phenomena. However, new problems arose with this conception because the structure of the molecules and how they are embedded in and affect the aether were unknown. Lorentz argued that it was best to accept the simplest assumption regarding the embedding: "except in the immediate neighbourhood of the particles – the properties of the aether are the same as in a vacuum."[149] However, he did not discuss the nature of the aether–particle interactions. Another interesting feature of this paper is that for the first time he presented a version of what were to become his "electrons." He hypothesized that bodies are systems of charged particles – tiny harmonic oscillators. He went on to argue that these charged harmonic oscillators within molecules are the cause of dispersion.

In his 1892 paper he was to develop the first full version of his "theory of electrons."[150] Before turning to that, though, we need to discuss his transition to a continuous-action conception of electromagnetic actions. In his work thus far, Lorentz had retained Helmholtz' version of electromagnetic actions. Thus, his conception of field at the start of his work was that it is a state of an aether, but that aether transmits action, between its particles, at a distance. Admittedly, this is a difficult conception to understand, and Lorentz never clarified it. In fact, a distinct feature of his work is that he never discussed the supposed structure of the aether, e.g., he never discussed how the aether particles and the matter particles were to interact. He simply asserted his agreement with Helmholtz' interpretation for the reason noted above. The properties of his own aether are mostly undefined.

Somewhere between 1878 and 1891 Lorentz adopted the "Faraday–Maxwell" continuous-action view. The first public announcement of the change was made in an address of 1891, "Electriciteit en Aether."[151] Most of his work during that period concerned the kinetic theory of gases, not electromagnetism; so it is difficult to say precisely when the change in position took place. Most likely it did not occur until after Hertz' discovery of electromagnetic waves in 1887. Although this discovery was not the only reason for the change, Lorentz did believe that, given Hertz' discovery, the old views could be saved only by artificial assumptions.[152] The other major reasons Lorentz gave were all related to the idea that with the continuous-action view the medium in the surrounding space plays a crucial role in understanding and accounting for electrostatic and electromagnetic energy, which is not possible with the action-at-a-distance account. He felt that the inclusion of the medium in such discussions accounted better for the observed phenomena.

It should be stressed here that although he changed his position with regard to the nature of the transmission of the actions, he remained committed to the existence of charged particles. They, in fact, were the focus of his conception. In his inaugural address of 1878 (as Full Professor of Theoretical Physics at Leiden at 24!), he said that he hoped that his research program into optical properties of matter based on the molecular theory and the electromagnetic theory of light would lead to an understanding of the molecular structure of matter. With regard to the molecular theory of matter, he said:

> Hardly anyone today will not know that physicists conceive of every body as a system of very small particles, the so-called *molecules*, each of which, as especially chemistry teaches us, may be composed of a number of smaller particles, the *atoms*.[153]

The support for this widespread belief in molecules came primarily from the successes of the kinetic theory of gases, to which Lorentz, himself, contributed significantly. Another important influence, for continental physicists, was the assumption that Weber had used to account for electromagnetic induction: Electric currents consist of positively and negatively charged particles, moving in opposite directions.

In ending his 1891 discussion of the reasons for the preference he was now giving the "Faraday–Maxwell" view, he noted the connection between the "old" theory (continental) and the "new" one (Maxwell):

> It seems to me not without interest to observe that, at least as far as form is concerned, one can, through a slight alteration, work out a rapprochement between the new conception and the old. In the earlier view, one transferred what one had observed of charged conductors to the particles of the imaginary electrical matter. A follower of Maxwell can proceed in a similar way. One can assume that there exist tiny charged particles, i.e., particles with properties similar to those of a charged conductor, and suppose that an observable charge – I mean the charge of a body of observable dimensions – consists in an accumulation of such particles and that an electric current consists in the motion of those particles.[154]

He began working out this *rapprochement* in his 1892 paper, i.e., he developed the initial version of his "theory of electrons." He proposed the following "physical hypotheses": (1) Material bodies are systems of microscopic charged "ions" (with assumptions about how they bind to make macroscopic bodies); (2) "Ions" are in part mechanical bodies to which Newton's laws apply; and (3) The aether is everywhere locally at rest. He had arrived at hypothesis (3) in an earlier work in which he had compared Stokes' hypothesis concerning the state of motion of the aether with that of Fresnel.[155] This issue will be discussed in the following section, after completing the present discussion of Lorentz' 1892 paper. One other important feature of this paper is that no assumptions about the structure of the aether were made, only of matter.

Given Lorentz' conception, it was necessary to reestablish a connection between matter and the aether, since bodies in motion may have currents and other electromagnetic phenomena taking place within them, and these are the result of the interaction between the "ions" and the aether. It was in the solution to this problem that the *rapprochement* was brought about: (1) Changes in states of the aether occur only by the presence and motion of "ions" and (2) "Ions" interact with one another only by changing the state of the aether, thus altering the force with which it acts on other "ions" at a *later* time. So, here we have both the charged particles of action at a distance and the delayed, contiguous actions of the continuous-action view. The 'ponderomotive' force with which the aether acts on the "ions" is a new force – the 'Lorentz force' – derived from the model by using what Lorentz called "mechanical" considerations.

Given the discussion of 'mechanical' and 'dynamical' in the previous chapter, we can see that, in our terms, Lorentz' attempted "mechanical" derivation of the electromagnetic equations was really 'dynamical'. That is, he applied the principles of generalized dynamics to the system of "ions" and the aether, and thus did not even attempt a 'mechanical' derivation in the sense Maxwell had desired.[156] Using general dynamical considerations, he was able to assume both of Maxwell's divergence equations and the equation for the circulation of the magnetic field. (It should be noted that the form of these equations had been simplified by Hertz and that they were now called the "Maxwell–Hertz equations.") He then derived the equation for the circulation of the electric field, which completed Maxwell's equations, and his own new force equation. After this paper he simply assumed all of the equations as the fundamental starting point for his theory.

The distinction between 'mechanical' and 'dynamical' is crucial here. In the previous discussion, we saw that 'dynamical' encompasses a wider range of mechanical systems than simply Newtonian. Lorentz' aether turns out to be a *non-mechanical, dynamical* system. That is, it is *non-Newtonian* in that it violates the fundamental law of action and reaction. With the Lorentz force, the aether acts on the "ions," but there is no mechanical reaction on the aether from the "ions." This is necessary in order to maintain the immobile aether hypothesis. Thus, for Lorentz, the electromagnetic field is a state of a non-Newtonian dynamical system. He explicitly acknowledged this in his 1895 paper:

> Admittedly this conception would violate the law of action and reaction – since we have reason to say that the aether *exerts* forces on ponderable matter – but, as far as I see, there is nothing to compel raising that law to a fundamental law of unlimited validity.[157]

That is, he had arrived at the conclusion that electromagnetic actions need not conform to the laws of classical mechanics. This was a "surprising and audacious step," but, characteristically, Lorentz did not comment much on it. In response to criticism by Poincaré he said, "But must we, in truth, worry ourselves about it [the violation of the third law]."[158] He maintained that it was a necessary consequence of retaining the hypothesis of an immobile aether, which he considered more important! We should now examine why he thought this hypothesis should be maintained despite the high cost, and what further consequences it had in Lorentz' work.

5.3 The immobile aether

I began the previous section by noting that Lorentz created his *rapprochement* while attempting to solve the central problems facing electromagnetism at the time. The first problem noted was that concerning the state of motion of the aether. Whether or not the luminiferous aether is in motion, or perhaps takes part in the motion of bodies which traverse it, especially the earth, had been a burning question in optics since the early 19th century. If one accepted the electromagnetic theory of light proposed by Maxwell, the problem became one for electromagnetism generally. That is, the electromagnetic fields produced in a moving aether would differ from those produced in a stationary aether. Surprisingly, Maxwell had not discussed this problem. One would think that, since *he* had proposed the identity of the luminiferous and the electromagnetic aethers, the existing literature on the optics of moving bodies would have been discussed at length by him. However, it is not so surprising if we recall the unclarity of his conception: Moving bodies are distinguished from the overall medium only by the values given to the electric and magnetic constants. For Lorentz, whose theory hinged on a clear distinction between matter and the aether, knowledge of the state of motion of the aether, itself, became essential.

At the time, only two major alternative hypotheses were widely accepted: those of Fresnel and Stokes. These were designed, in particular, to account for the phenomenon of stellar aberration. Let me briefly fill in the background of these hypotheses.[159] In 1728, the astronomer Bradley had reported that starlight observed through a telescope appears displaced by a small angle. This "aberration" requires that telescopes be sighted parallel to the apparent path of the light, rather than to the actual path. He calculated the "angle of aberration" on the basis of the relationship between the velocity of the earth and the velocity of light. This phenomenon was of great interest to scientists because it provided observational evidence of the motion of the earth. Aberration could be explained, rather convincingly, on the basis of the corpuscular theory of light by using the classical addition of velocities theorem and the assumption that light particles are not affected by the earth's gravitation. However, it presented more severe difficulties for the wave theory of light, since aberration seems to imply that the aether does not take part in the motion of the earth. Young made the suggestion that the aether is stationary, permeates all matter, and offers no resistance to moving matter. Fresnel was to develop this suggestion fully.

Fresnel proposed, based on the corpuscular theory, that there should be an additional aberration effect if starlight were to be passed through a refracting medium. Arago conducted such experiments and found no effect. Fresnel found that it was easier to account for Arago's results with the wave theory. He formulated (1818) a mathematically simple and elegant way of accounting for all aberration effects, based on the following hypotheses: (1) The aether permeates all matter; (2) It has no overall motion in the direction of the earth's motion; and (3) There is some local motion due to the density of the aether being greater in matter than in a vacuum, i.e., the *excess* aether in moving matter is *dragged* along with the matter. He concluded that the velocity of light in moving media would be less than was to be expected on the basis of the classical addition of velocities theorem. The amount by which it is less came to be known as the "dragging coefficient" or the "coefficient of entrainment."[160] Fresnel's hypothesis accounted for all the known effects and subsequently received much experimental support; in particular, from the experiments of Fizeau on the velocity of light in moving refractive media.

However, in 1845, Stokes showed that it is possible to assume that the earth drags the aether completely along with it and still get the same agreement with observation. He hypothesized that: (1) The aether near the surface of the earth is set in motion by the earth: (2) The velocity of the aether at any point on the surface of the earth is equal to the velocity of the earth; and (3) The motion of the aether is irrotational.

Mathematically, both hypotheses seemed to provide the same results. Observationally, it seemed nearly impossible to detect any motion of the aether with respect to the earth, since a highly sensitive, second-order (v^2/c^2) experiment would be required. Interestingly, Maxwell, in his encyclopedia article on the aether, had suggested such an experiment but dismissed it since the expected result would be "quite insensible."[161] Michelson picked up this suggestion and attempted the first "aether-drift" experiment in 1881. These experiments were continued until 1930 (!) in collaboration with Morley and with Miller.[162] Thus, the famous Michelson–Morley experiment of 1887 was designed as a 'crucial experiment' between *rival aether hypotheses* and not as a test for *the existence of the aether*, itself, as it is sometimes stated in physics texts. The null result did not automatically lead to special relativity and the abandonment of the aether, nor did it lead to the automatic acceptance of the hypothesis it seemed to confirm – that of Stokes. In the hands of Lorentz, it led, instead, to the retention of the Fresnel hypothesis of a stationary aether by the introduction of an "audacious" new hypothesis.

Lorentz first discussed the two rival aether hypotheses in 1886.[163] At that time, he felt he had conclusively shown Stokes' hypothesis to be untenable because his three conditions are inconsistent.[164] He also criticized Michelson's 1881 experiment, showing that because of errors in calculation and design the null result could not be taken as established. Additionally, he considered a combined Fresnel–Stokes hypothesis but did not retain it because it had no real advantages over the simpler Fresnel hypothesis. He was to change the character of this hypothesis significantly in 1892.

The stated purpose of Lorentz' 1892(a) paper was "to know the laws which govern the movement of electricity in bodies which traverse it without carrying it along."[165] He achieved this by using Maxwell's laws as the foundation and making the hypotheses concerning matter and the aether we have discussed in the previous section. Lorentz' 1892 aether is the stationary aether of Fresnel with one crucial difference: Lorentz' aether is *everywhere locally at rest.* That is, he was able to provide a "theoretical deduction" of the "dragging coefficient" without Fresnel's hypothesis of the actual dragging of the aether. He interpreted the effect as the result of interaction between light and the charged harmonic oscillators. On Lorentz' picture, the reduction in the expected velocity of light occurs because the incident light waves shake the "ions," creating new waves which interfere with them.[166] He was to discard this assumption as unnecessary in his 1895 paper.[167]

It is rather curious, given that Michelson had repeated his experiment with Morley in 1887 and had obtained what was considered to be a conclusive null effect, that Lorentz did not discuss the fact that this result contradicts his immobile aether hypothesis. He only discussed first-order effects. However, the Michelson–Morley experiment was certainly on his mind, and, shortly after writing the 1892 paper, he wrote a letter to Rayleigh expressing his concern, which I will quote here at length:

> Fresnel's hypothesis taken jointly with his coefficient $1-1/n^2$ would serve admirably to account for all the observed phenomena were it not for the interferential experiment of Mr. Michelson, which has, as you know, been repeated after I published my remarks on its original form, and which seems decidedly to contradict Fresnel's views.
>
> I am totally at a loss to clear away this contradiction, and yet I believe if we were to abandon Fresnel's theory we should have no adequate theory of aberration at all, the conditions which Mr. Stokes has imposed on the movement of the aether being irreconcilable to each other.

> Can there be some point in the theory of Mr. Michelson's experiment which has as yet been overlooked?
>
> In the meantime I have endeavoured to apply the electromagnetic theory of light to a body which moves through the aether without dragging the medium along with it; my paper is now under the press and I hope, in a few weeks, to be able to offer you a copy of it. Assuming a supposition which may appear somewhat startling but which may, as I think, serve as a working hypothesis, I have found the right value . . . for Fresnel's constant.[168]

The reference to a "startling supposition" must be to the hypothesis concerning the "vibrating ions," used to explain the dragging effect. Later in the year, he found another hypothesis (which must have been even more "startling"!) to account for the apparent contradiction: the change in the dimensions of matter in motion through the aether. Since the way in which he introduced and discussed that hypothesis reveals interesting features about Lorentz' conception and about his way of thinking, I will again quote him at length:

> This experiment [Michelson–Morley] has been puzzling me for a long time, and in the end I have been able to think of only one means of reconciling its result with Fresnel's theory. It consists in the supposition that the line joining two points of a solid body, if at first parallel to the direction of the earth's motion, does not keep the same length when it is subsequently turned through 90° . . . .
>
> Now, some such change in the length of the arms in Michelson's first experiment and in the dimensions of the slab in the second one is, so far as I can see, not inconceivable. What determines the size and shape of a solid body? Evidently the intensity of the molecular forces; any cause which would alter the latter would also influence the shape and dimensions. Nowadays we may safely assume that electric and magnetic forces act by means of the intervention of the ether. It is not far-fetched to suppose the same to be true of the molecular forces. But then it may make all the difference whether the line joining two material particles shifting together through the ether, lies parallel or crosswise to the direction of that shift . . . .
>
> Since the nature of the molecular forces is entirely unknown to us, it is impossible to test the hypothesis. We can only calculate – with the aid of more or less plausible suppositions, of course – the influence of the motion of ponderable matter on electric and magnetic forces . . . .
>
> One may not of course attach too much importance to this result [his calculation of the contraction of matter in the direction of the earth's

motion]; the application to the molecular forces of what was found to hold for electric forces is too venturesome for that. Besides, even if one would do so, the question would remain whether the earth's motion shortens the dimensions in one direction as assumed here, or lengthens those in directions perpendicular to the first, which would answer the purpose equally well.[169]

What we see from this discussion is the following. First, Lorentz introduced the 'contraction hypothesis' completely *ad hoc.*[170] It was designed to account for this *one* experiment.[171] Second, the 'contraction hypothesis' was really a more general hypothesis concerning the *change of dimensions* of matter in motion through the aether: possibly a contraction in the direction of the earth's motion or, *equally possible*, a lengthening in the perpendicular direction. Lorentz' decision to pursue the contraction possibility was completely arbitrary. He calculated the approximate value of the contraction assuming that molecular forces transform like electric and magnetic forces; in particular, like the Lorentz force, derived in his earlier paper. Third, he was willing to accept an hypothesis which he considered to be untestable for the foreseeable future. Finally, we see that, for Lorentz, the stationary aether hypothesis was so necessary that he was willing to make a rather bizarre assumption in order to retain it – an assumption which *he* considered much more bizarre than *simply* denying the universal validity of the Newtonian law of action and reaction! He was, in fact, willing to assume a *causal* relationship between the dimensions of moving matter and the aether.

Lorentz brought together, in a systematic and coherent fashion, all of the elements of his work of 1892 in his highly influential paper of 1895.[172] The important features of this work, with respect to our discussions so far, are: (1) He explicitly acknowledged and discussed the non-Newtonian nature of his aether and (2) He discussed the Michelson–Morley experiment, first by analyzing it in accordance with the classical addition of velocities theorem and then showing how his hypothesis concerning molecular forces could provide for the observed null effect.

Additionally, Lorentz formulated a "theorem of corresponding states" which provided a way of transforming the electromagnetic equations from a moving system to one at rest with respect to the aether, which made the laws of electromagnetism partially covariant. I will now discuss these 'Lorentz transformations' insofar as is necessary for an understanding of the contention I wish to uphold in the final section of this chapter, namely that Lorentz' conception of field remained that of a state of the aether and not of "mere space."[173]

5.4 Lorentz' interpretation of the 'Lorentz transformations'

The first time Lorentz introduced a modification to the standard ('Galilean') transformations of Newtonian mechanics was in his 1892(a) paper. In order to transform the wave equation for radiation from a rest system to a moving system and maintain the correct form, an additional mathematical transformation was needed. Both the spatial translation in the direction of motion and the temporal transformation were altered. However, Lorentz considered this additional transformation to be a purely mathematical coordinate transformation, like that from Cartesian to polar coordinates. He went on to formulate a "general theorem": The electromagnetic field equations, with respect to first-order phenomena, have the same form in a rest (aether) system and in a system moving with constant velocity relative to it. This required that the spatial transformations retain their Galilean form, but the temporal one is 'mixed', i.e., it contains spatial coordinates as well.

In his 1895 paper Lorentz produced two sets of altered transformation rules. First, in considering the problem of transforming an electrodynamic system to an electrostatic one (i.e., from a moving system to one at rest in the aether), he had to alter the transformation rule for the spatial coordinate in the direction of the earth's motion. This change has the effect of 'stretching' and 'shrinking' that dimension. That is, when making a transformation from a moving system to a system at rest, the X-dimension would *increase*, while a transformation from a system at rest to a moving one would produce a *decrease* in dimension. Again, at this point in his work, Lorentz considered this to be a purely mathematical mapping and not a physical alteration. The 'ions' were defined as rigid. As we will see, it was not until 1899 that Lorentz was to make an explicit connection between the altered spatial transformation for electric and magnetic forces and his 'contraction hypothesis' (introduced in 1892) for molecular forces.

The second set of altered transformations arose in transforming problems regarding optical phenomena in moving bodies into problems in stationary ones. Here, the spatial transformations remained unaltered, but the temporal one had to be changed in the same manner as in the previous paper. Lorentz called the time "local" ("*Ortszeit*") in the moving system and "general" ("*allgemeine Zeit*") in the rest system. As before, this alteration was simply mathematically expedient. He reestablished his "general theorem" as the "theorem of corresponding states" using only the altered time transformation. The last section of the paper contained the discussion of the 'contraction hypothesis' noted previously but did not relate this to the altered Galilean transformations of the preceding sections. These, and his theorem, were only valid for first-order phenomena.

Lorentz produced his theorem in a "less troublesome way" in his 1899 paper, this time using both alterations to the Galilean transformations.[174] It was still only applied to first-order phenomena, where one can make the "simplest assumption" that "molecular forces . . . are *not* changed by the translation of the system."[175] However, this paper was more than a "simplification" of his previous work. He added a proposed set of second-order transformation rules but was dissatisfied with these because of the presence of an "indeterminate" factor. It is with these transformations that he first linked the 'contraction hypothesis' with the altered X-coordinate. For second-order phenomena he assumed that "the transformation *really* takes place, of itself, i.e., by the action of the forces acting between the particles of the system and the aether."[176] He went on to say: "The transformation of which I have now spoken is precisely such a one as is required in my explanation of Michelson's experiment."[177] From this point on Lorentz was to argue that a *real* change in dimension is reflected in the new spatial transformation rule but that the new temporal rule has no physical significance. Additionally, he proposed an hypothesis, which he again called "startling," in considering the ratio of the mass of an "ion" in a moving system with its mass in a stationary system. Given his transformations, the value could not be unity:

> . . . consequently, states of motion, related to each other in the way we have indicated, will only be possible if in the transformation . . . the masses of the ions change; even this must take place in such a way that the same ion will have different masses for vibrations parallel and perpendicular to the velocity of transmission.[178]

He went on to argue for the plausibility of such an assumption and concluded that, given it, one would expect the negative result of the Michelson–Morley experiment.

Lorentz produced the complete version of what Poincaré was to call the "Lorentz transformations" in his paper of 1904.[179] He succeeded in accounting for all known electromagnetic phenomena by making, in total, eleven assumptions (!); in particular, the transformation rules which make the equations of electromagnetism invariant. His 'contraction hypothesis' now had the form that the shape of the individual "electron" (discovered by J. J. Thomson in 1897) is altered during motion, thus altering the dimensions of moving bodies. Except for some mathematical errors, which made him restrict his theory more than was necessary, Lorentz had now finally produced his "theory of electrons." The results of this paper are formally equivalent to those obtained by Einstein in his special relativity paper of 1905. However, Lorentz' interpretation differs substantially from Einstein's, which was proposed just one year later and which has since become customary, in that: (1) The transformation rules are non-reciprocal, e.g., inverse spatial transformations produce a *dilation* effect, rather than a contraction; (2) The change in dimension appearing in the X-coordinate transformation represents an *actual* change in the dimensions of moving bodies; and (3) The temporal transformation represents the "local" time in the moving system and is simply a *mathematical* coordinate transformation. Lorentz' interpretation is, thus, asymmetrical: One altered transformation represents a real physical change, while the other only represents a mathematical convenience. He did not seem concerned by this. As far as I can find, he discussed the peculiar asymmetry of his interpretation only once – in a letter to Einstein in 1915, which will be discussed in the next section.

5.5 Summary: L'éther pour toujours

The historian Hirosige, who has written the most comprehensive study of Lorentz' contributions to the development of the electromagnetic field concept thus far, has concluded that as of 1895:

> . . . we can see the modern concept of the electromagnetic field definitively settled. Though Lorentz still uses the expression 'a state of the aether' instead of 'a state of the electromagnetic field,' his aether is no longer a dielectric medium conceived in analogy with ponderable matter; it is an independent reality which supports all electromagnetic action. Apart from the name, it is equivalent to our notion of electromagnetic field.[180]

What is correct in what Hirosige is saying is that Lorentz' aether functions in the way we would today say the field does. However, for Lorentz, there was more at stake than a "name." There is also more at stake for the analysis presented here, since I am attempting to show how concepts develop in scientific practice. According to my analysis, Lorentz began with Helmholtz' conception of the electromagnetic field as a state of an action-at-a-distance aether and ended wth his own conception (as of 1892) that the electromagnetic field is a state of a non-Newtonian aether, i.e., one which violates the law of action and reaction. This is not the "modern concept" – which should not be attributed to anyone before Einstein – that the electromagnetic field is a state of space and ontologically on a par with matter.

For Lorentz, the field is not an "independent reality" (which I interpret to mean 'ontologically on a par with matter') since it was always essential for his theory that space be filled with a substantial aether and that the field be a *state* of that aether. The aether may be an "independent reality," but then it is not equivalent to the "modern concept of field." It is true that Lorentz never concerned himself with the *structure* of the aether. He never attempted to give a 'mechanical' explanation of it in the way Maxwell and others tried, and he gave up any attempt at a 'dynamical' explanation after his analysis of 1892. He simply assumed the field equations as fundamental. The properties of his aether remained mostly undefined.

However, the aether was so important to his conception that he retained it even after special relativity was distinguished clearly from his theory of electrons, after the development of the general theory of relativity, and even after the development of quantum mechanics ended the viability of his theory. I will first present some of the evidence that this is so and then discuss why. In the process I will summarize the major points which have been made concerning Lorentz' conception.

The 1909 edition of Lorentz' *Theory of Electrons* contains his first response to Einstein's special relativity:

> His [Einstein's] results concerning electromagnetic and optical phenomena . . . agree in the main with those which we have obtained in the preceding pages, the chief difference being that Einstein simply postulates what we have deduced, with some difficulty and not altogether satisfactorily, from the fundamental equations of the electromagnetic field. By doing so, he may certainly take credit for making us see in the negative result of experiments like those of Michelson, Rayleigh and Brace, not a fortuitous compensation of opposing effects, but the manifestation of a general and fundamental principle.
>
> Yet, I think, something may also be claimed in favour of the form in which I have presented the theory. I cannot but regard the ether, which can be the seat of an electromagnetic field with its energy and its vibrations, as endowed with a certain degree of substantiality, however different it may be from all ordinary matter. In this line of thought, it seems natural not to assume at starting that it can never make a difference whether a body moves through the ether or not, and to measure distances and lengths of time by means of rods and clocks having a fixed position relatively to the ether.
>
> It would be unjust not to add that, besides the fascinating boldness of its starting point Einstein's theory has another marked advantage over mine. Whereas I have not been able to obtain for the equations referred to moving axes *exactly* the same form as those which apply to a stationary system, Einstein has accomplished this by means of a system of new variables slightly different from those which I have obtained. I have not availed myself of his substitutions, only because the formulae are rather complicated and look somewhat artificial, unless one deduces them from the principle of relativity itself.[181]

Thus, although Lorentz admitted that Einstein had achieved the same, and in some ways better, results by simply assuming invariance at the outset and thus eliminating the preferential status of the aether frame, he continued to retain his own interpretation of the transformations and to express his fundamental belief in the existence of the aether. His comments are interesting both because of this and because they express his fundamental disagreement with Einstein's way of doing physics (a fascinating subject in its own right to which we will return later).

In a 1917 lecture on Einstein's general relativity, Lorentz discussed the question of what should serve as the reference system for electromagnetic phenomena which take place in space and do not take part in the earth's rotation. He suggested that the aether could serve that function well:

> Would it not be quite right to imagine a stationary aether, with respect to which the earth rotates and in which electromagnetic waves propagate without being disturbed by the motions of the earth?[182]

He went on to argue that:

> It is always risky to close a path of research completely and perhaps it is good, everything taken together, to grant the aether one more chance. Conceivably a time will come when speculations over its structure, from which we now abstain, become fruitful and effective.[183]

He seems to be saying here, and in other articles in this vein, that the *conceptual* difficulties of doing without the aether are so great, perhaps we should retain it in some form. Before discussing what form, let me just add a few more pieces of evidence to support my claim that Lorentz remained committed to the existence of an aether to the end of his life.

First, in his 1922 lectures to the California Institute of Technology, he began by begging "permission" from the students to "use the time-honored word, although nowadays some [!] physicists prefer not to speak of an aether."[184] But he did not mean simply to use it as a substitute for the word 'vacuum'. He qualified his use only by saying that "we must be careful not to assign *so much of the properties* of ordinary matter as was done in old times [italics mine]."[185] Second, in a 1923 article, Lorentz still listed the detection of the earth's motion with respect to the aether as a possible way of measuring its rotation.[186] Finally, there is some circumstantial evidence from the last year of his life in that he thought his 1901–2 lectures on aether theories to be of sufficient value to be published without alteration.[187]

Now, why did Lorentz retain his belief in the existence of the aether? First of all, Lorentz needed the aether as long as his theory of electrons was viable (approximately 1910), since the aether has a *causal relationship* to the size and mass of moving bodies. As discussed previously, Lorentz held that the electrons undergo an actual physical contraction due to the circumstances of their motion through the aether. Although his initial arguments for the proposed contraction were *ad hoc* and geometrical, it became a central hypothesis of his theory that the contraction occurs because the molecular forces in matter are affected by their motion through the aether, as are electric and magnetic forces. A further consequence of his theory is that the mass of an electron is dependent upon its motion through the aether. Thus, the aether was essential to Lorentz' theory because motion through it is the cause of postulated physical phenomena, for which there was evidential support.

However, even after his theory was no longer viable, i.e., even after the postulated causal interaction no longer had a basis in theory, Lorentz retained the aether.[188] I will conclude this discussion of Lorentz by showing that, despite the many radical aspects of Lorentz' theory and methods, with regard to the notions of space and time he remained a conservative.[189] *Lorentz never stopped believing that the aether could function as an 'absolute' reference frame for all the motions in the universe.*

Lorentz had established the aether as the universal reference frame in his 1895 paper. There he maintained:

> That there can be no talk of *absolute* rest of the aether is self-evident; the expression would not even make sense. When for brevity I say that the aether is at rest, I mean only that there is no dislocation of one part of this medium with respect to the others and that all observable motions of celestial bodies are relative motions with respect to the aether.[190]

Thus, although not the Newtonian absolute reference frame, the aether functions as an absolute reference system in that it is at rest with respect to all other reference systems. It seems that by 1910 Lorentz had come to the conclusion: "Finally, there is only so much substantiality left to it [the aether] that through it one can determine a reference system."[191] As was discussed in the preceding section, Lorentz' interpretation of his transformation rules requires the aether. That is, the aether is the reference system with respect to which it is possible to determine true, universal time and true length. Additionally, the aether frame, where clocks and rods are at rest with respect to the aether, is the only frame where the velocity of light is *truly* constant – it only *appears* constant in other frames because of the contraction effect.[192]

What bothered him most in relativity theory was the difficulty in conceptualizing 'space' and 'time'. In 1915 he said that he believed that "one can accept the essence of the relativity principle, without breaking with the old concepts of space and time."[193] He was concerned particularly with the concept of time and, in several places, discussed the "paradox" that, given Einstein's interpretation, it is not possible to determine which observer, moving or stationary, possesses the correct time. He said of this "paradox" in his 1922 lectures:

> A physicist of the old school says, 'I prefer the time that is measured by a clock that is stationary in the ether, and I consider this as the true time, though I admit I cannot make out which of the two times is the right one, that of A or that of B.' The relativist, however, maintains that there cannot be the least question of one time being better than the other.
>
> Of course this is a subject that we might discuss for a long time. Let me say only this: All our theories help us to form pictures, or images, of the world around us, and we try to do this in such a way that the phenomena may be coördinated as well as possible, and that we may see clearly the way in which they are connected. Now in forming these images we can use the notions of space and time that have always been familiar to us, and which I, for my part, consider perfectly clear and, moreover, as distinct from one another. My notion of time is so definite that I clearly distinguish in my picture what is simultaneous and what is not.[194]

Thus, we see that despite Einstein's critical analysis of the concepts of space and time in the special theory and despite his geometrization of space–time in the general theory, Lorentz retained the "old" notions. Although he expressed his gratitude to Einstein for his theory, which would not have been possible if he had "gone along old fashioned roads" and retained the aether, Lorentz could never fully accept its consequences.

In 1915, Lorentz wrote a very interesting and, because of its 'speculative' content, rather uncharacteristic letter to Einstein.

This letter highlights some important points which have been made in my analysis. Lorentz was objecting to Einstein's claim, in his *Die Kultur der Gegenwart* article on relativity, that he had looked "in vain for sufficient reason" why for the formulation of the laws of physics a system at rest is preferable to a rotating system and that he had thus "felt compelled rather to postulate that both systems are equally justifiable."[195] Against this, Lorentz said he could find "sufficient reason" for the distinction – "the two systems move in different ways with respect to the aether."[196] He acknowledged that, given Einstein's denial of the existence of the aether, his claim was correct. But, Lorentz cautioned, "[d]o you not go somewhat too far here, in that you put forward a personal view as self evident?"[197]

In discussing the 'contraction hypothesis', he admitted that he had introduced it *ad hoc* and said of his supposition concerning the alteration of molecular forces: "I must, indeed, confess that I first made these remarks *after* I had found the hypothesis."[198] He went on to make remarks of a "more didactic nature" concerning the contraction, space, and time:

> If one derives the 'shortening' from the equations of relativity theory (which by itself is of course entirely justified) and nothing further is added for clarification, one risks creating the impression that the question here is about 'apparent' things rather than about a real physical phenomenon; at least I have found from time to time expressions by representatives of relativity theory which appear to give evidence of such an interpretation. As against this one may remark that when we observe an 'alteration' . . . according to the customary linguistic usage (and why should we not cling to that?), this 'alteration' represents a physical phenomenon. The shortening of a fixed rod, that is made to move with respect to K, is just as real as the expansion by raising the temperature . . . .[199]

Also, in discussing simultaneity:

> We perceive entirely clearly not only 'next', 'behind', and 'above', but also 'after' one another. But to me it appears that there is an unmistakable difference between the ideas of space and time, a difference which you cannot entirely put aside. You cannot consider the time coordinate as fully equivalent with the space coordinate . . . . As far as time is concerned, thus we have, as it seems to me, a completely clear conception of what successive moments are, and also of 'simultaneity'.[200]

Finally, he argued that it is entirely possible to determine correct time if one would "decorate or should I say disfigure" one of the systems with an aether with respect to which clocks would be at rest, as he preferred to do.[201]

What we see most clearly from this letter is that Lorentz' discomfort with Einstein's elimination of the aether from the field concept had its roots in his reluctance to give up the possibility of a reference frame with respect to which we can determine, in principle if not in practice, the *real* time of events and the *real* dimensions of objects. Such a possibility fits better with our 'intuitive' notions of 'space' and 'time', which have been conditioned by our experiences. What Einstein had argued is that if we reflect on what we mean by 'space' and 'time' we will see that our 'intuitive' notions are confused and unfounded. It *was* "only a short step" to the conception of the electromagnetic field as a state of space, but what is required was precisely the type of critical, 'philosophical' analysis Lorentz was reluctant, or unable, to make. Nevertheless, his conception of the field as a state of a non-Newtonian aether was a significant change in the concept and was influential on Einstein's thinking. Einstein, near the end of his life, gave his own assessment of Lorentz' contributions to "our modern conceptions":

> People do not realize how great was the influence of Lorentz on the development of physics. We cannot imagine how it would have gone had not Lorentz made so many great contributions.[202]

Albert Einstein (1897–1955)

CHAPTER 6

Einstein's 'field'

6.1 Critical reflections

"The introduction of a 'luminiferous' ether will prove to be superfluous . . . ".[203] As far as the question of the meaning of 'electromagnetic field' is concerned, we can stop here – with the third paragraph of Einstein's 1905 paper on relativity. If the aether is "superfluous," the field must be a state of space itself, which is the present conception. However, since we are concerned with the *development* of the field concept, we need to discuss what led Einstein to dismiss, in one bold stroke, the major object of concern of late-19th century physicists.

What was it that led Einstein to assume as a principle, the very thing that Lorentz had "with difficulty and not altogether satisfactorily" deduced: the validity of the laws of electrodynamics in all inertial frames? It is not possible to answer this by reading the paper itself. There is no discussion of the question in the paper and there are no references to the work of other scientists; not even to Lorentz. Yet, as we have already seen to some extent, Einstein clearly saw himself as continuing the work of Lorentz and he was to claim repeatedly that relativity theory was the culmination of the "Faraday–Maxwell epoch."

For insight into how he arrived at his 1905 presentation, we will have to rely on his later reflections; in particular, on his "Autobiographical notes."[204] Now, it is true that there is a danger in so doing. Einstein, himself, cautioned in his later years:

> If you want to find out anything from the theoretical physicists about the methods they use, I advise you to stick closely to one principle: don't

> listen to their words, fix your attention on their deeds. To him who is a discoverer in this field, the products of his imagination appear so necessary and natural that he regards them, and would like to have them regarded by others, not as creations of thought but as given realities.[205]

However, recent works by several historians and scientists substantiate what Einstein said in his "notes" sufficiently to consider them a reliable source of information.[206] I will use material from these works in my analysis. Additionally, I will make use of some of Einstein's "deeds": the papers of his *annus mirabilis* (1905), a review paper of 1907, and some later papers.[207] Although there are instances in which Einstein's 'caution' should be used with him, his acute ability to scrutinize "the products of his imagination" makes him one of the best sources concerning their roots.

In all, special relativity was the product of a ten-year process for Einstein. He, himself, traced its origin to a thought–experiment of 1895 and 1896, when he was a high school student in Aarau:

> If I pursue a beam of light with the velocity c (velocity of light in a vacuum), I should observe such a beam of light as a spatially oscillatory electromagnetic field at rest. However, there seems to be no such thing, whether on the basis of experience or according to Maxwell's equations.[208]

It is interesting to note, as has been pointed out by several authors, that there is no such problem if the classical (Galilean) transformation rules are used with Maxwell's equations. But Einstein seems to have assumed from the outset that different transformations would be necessary:

> From the very beginning it appeared to me intuitively clear that, judged from the standpoint of such an observer, everything would have to happen according to the same laws as for an observer who, relative to the earth, was at rest.[209]

He went on to say that "one sees that in this paradox the germ of the special relativity theory is already contained."[210]

At this point, however, Einstein still believed in the existence of the aether. His first 'paper', sent with a letter to his uncle in 1894 or 1895, was on the subject of the "state of the aether in the magnetic field."[211] Although it in no way presages later developments, it is interesting because it presents Einstein's "program" of study at the time: to determine the magnetic field produced by an electric current, by measuring the elastic deformation of the aether. It is not known when Einstein stopped believing in the aether's existence. He maintained that he had formulated new experimental methods for measuring the aether-drift while he was a student at the ETH in Zürich, but that he had failed to carry out the experiments. The last known evidence is in a letter to his friend, Marcel Grossman, in 1901. There he said that he had "a new and simple method for investigating the motion of matter relative to the light-aether."[212]

The *exact* date of his dismissal of the aether is not important. Einstein was led to eliminate the aether from the concept of field through a process of critical reflection on and reassessment of the foundations upon which the aether, and thus the field, rested. He began with an attitude which he maintained throughout his life: that fundamental problems in science require *epistemological* analysis and that science without such analysis is "primitive and muddled" – quite the antithetical view to that of Lorentz![213] Things were certainly muddled when Einstein began his study of physics.

On the one hand, in the attempts to reduce electromagnetism to mechanics:

> One got used to operating with these fields [electric and magnetic] as independent substances without finding it necessary to give oneself an account of their mechanical nature; thus mechanics as the basis of physics was being abandoned, almost unnoticeably, because its adaptability to the facts presented itself as finally hopeless.[214]

We noted this 'abandoning' in the transition from 'mechanical' to 'dynamical' explanations and in the ultimate assumption of the validity of the laws of electromagnetism, regardless of their underlying explanation. However, explicit acknowledgement that the desired reduction was "hopeless" could not be made because:

> . . . dogmatic rigidity prevailed in matters of principles: In the beginning (if there was such a thing) God created Newton's laws of motion together with the necessary masses and forces. That is all: everything beyond this follows from the development of appropriate mathematical methods of deduction.[215]

On the other hand, in the attempts to reduce mechanics to electromagnetism, it was necessary to retain the aether since all mechanical phenomena were to be reduced to states of the electromagnetic aether. Such a reduction would eliminate the "disturbing dualism" of particle and field, created by the work of Lorentz:

> . . . one could hope to deduce the concept of mass-point together with the equations of motion of the particles from the field equations . . . .[216]

This has not been achieved because:

> . . . no method existed by which this kind of field equations could be discovered without deteriorating into adventurous arbitrariness.[217]

Despite this:

> . . . one could believe that it would be possible by and by to find a new and secure foundation for all of physics upon the path which had been so successfully begun by Faraday and Maxwell.[218]

However, Einstein knew as early as 1900–1901 that this approach would not work:

> . . . a second fundamental crisis set in, the seriousness of which was suddenly recognized due to Max Planck's investigations into heat radiation.[219]

It should be noted that the "crisis . . . was suddenly recognized" *by Einstein* and not by the physics community as a whole. The significance of Planck's paper on black-body radiation (a problem in classical thermodynamics) went largely unnoticed. Indeed, Einstein seems to have been alone in his recognition of the full significance of Planck's quantization of the energy of radiation.[220] Einstein's realization of its significance led to his first two papers of 1905 – the first on the photo-electric effect and the second on Brownian motion – *and* to the relativity paper.

Einstein saw that both mechanics and electrodynamics are inadequate in regions small enough for fluctuation phenomena to count. "The contradiction with dynamics was here fundamental; whereas the contradiction with electrodynamics could be less fundamental."[221] That is, if radiation possesses discrete energy, it is in contradiction with mechanics. If it possesses an atomic structure it is in contradiction with mechanics and with electromagnetism. He saw the possibility of reconciliation with electromagnetism. Additionally, he saw that Lorentz' electron theory led to the prediction of the wrong amount of energy for black-body radiation. These conflicts between the various domains of physics were so apparent to Einstein because he saw physics as one unified discipline.

His skepticism concerning the then-accepted foundations of physics had its roots in his reading of the philosophical views of David Hume and Ernst Mach.[222] The former's views on space, time, and causality were to influence his analysis of simultaneity in the relativity paper; while the latter's view, that no scientific principles can be regarded as established truths but that they are always subject to the control of experience and are thus alterable, "shook his dogmatic faith" in the principles of Newton.

The skepticism awakened by Hume and Mach was reinforced by the recognition of the inconsistencies noted above. He was then led to attempt to formulate a new foundation for physics. Einstein began with a consideration of electromagnetic and mechanical phenomena on a macroscopic level. However, he found that all his attempts:

> . . . to adapt the theoretical foundations of physics to this knowledge failed completely. It was as if the ground had been pulled out from under one, with no firm foundation to be seen anywhere, upon which one could have built.[223]

From these attempts he "despaired of the possibility of discovery of the true laws by means of constructive efforts based on known facts."[224] Thus, the stage was set for a more daring approach:

> The longer and more despairingly I tried, the more I came to the conviction that only the discovery of a universal formal principle could lead to assured results.[225]

Here we see the fundamental difference between Lorentz' approach to physics and that of Einstein most acutely: For Einstein, "theories of construction," such as that of Lorentz, cannot resolve foundational problems. Resolution of such problems can only be achieved by "theories of principle."[226] Thus, while Lorentz assumed the transformation rules he had previously "constructed" and attempted to derive the invariance of the laws of electrodynamics from these, Einstein assumed the invariance of the laws and derived the transformations from this.

Einstein used another great "theory of principle" as a model for his relativity paper: classical thermodynamics and the impossibility of a *perpetuum mobile.* He assumed what seemed the obvious principles needed to clear up the "muddle." The first postulate – "for every reference system in which the laws of mechanics are valid, the laws of electrodynamics and optics are also valid" – eliminates the need for the aether as a preferred reference system for the formulation of the laws of electromagnetic theory.[227] Thus, he considered electrodynamics and mechanics to be on equal footing: neither was to be reduced to the other. The second postulate – "light is always propagated in empty space with a definite velocity *c* which is independent of the state of motion of the emitting body" – eliminates the role of the aether as the only frame in which the velocity of light is truly *c* (recall that for Lorentz it only appears to be *c* in all inertial frames because of the contraction effect).[228] Given these two postulates, the aether is of course "superfluous":

> These two postulates suffice in order to obtain a simple and consistent theory of the electrodynamics of moving bodies taking as a basis Maxwell's theory for bodies at rest. The introduction of a 'luminiferous ether' will prove to be superfluous because the view here to be developed will introduce neither an 'absolutely resting space' provided with special properties, nor associate a velocity-vector with a point of empty space in which electromagnetic processes occur.[229]

However, a problem immediately arises in that the two postulates appear contradictory. That is, the velocity of light should be different in systems in relative motion to one another *if* we use the Galilean transformation rules. Einstein realized that in order to make the two assumptions consistent an analysis of space and time was necessary. He later claimed that 'time' was the key to the resolution of the apparent conflict: Lorentz' "local time" should be taken as simply 'time', i.e., there can be no such thing as absolute or "general time."[230]

Einstein claimed that he was aided in this realization by his reading of Mach and Hume:

> The type of critical reasoning which was required for the discovery of this central point was decisively furthered, in my case, especially by the reading of David Hume and Ernst Mach's philosophical writings.[231]

Clearly, Mach's view that all scientific concepts are subject to change and his views on the *relative* nature of all motion must have been influential in Einstein's thinking. In a letter of 1948 to his friend Michele Besso, Einstein said that he felt that Hume's influence on his thinking had been greater.[232] Although it certainly is not possible to make clear correlations between Einstein's analysis of 'space' and 'time' and Hume's discussion of these concepts, there are certain key statements in Hume's *Treatise* which Einstein's analysis brings to mind:

> . . . *the idea of space or extension is nothing but the idea of visible or tangible points distributed in a certain order* . . . [italics in original][233]

> . . . we have no idea of any real extension without filling it with sensible objects . . . that we have no such idea ["the idea of time without any changeable existence"] is certain.[234]

> . . . time cannot make its appearance to the mind, either alone, or attended with a steady unchangeable object, but is always discovered by some *perceivable* succession of changeable objects.[235]

6.2 'Electromagnetic field' in the special theory

Einstein began his paper with an analysis of the kinematics of moving bodies. All electrodynamics must be based ultimately on kinematics since all measurements of electromagnetic phenomena are made by means of 'rigid' bodies and clocks. Einstein maintained that the problems which arose in interpreting experimental results came about because close enough attention was not given to the kinematics of moving bodies. It should be noted here that he was concerned only with *first-order* experiments, e.g., aberration, Fizeau's experiments, etc., and not with those of second-order, in particular, the Michelson–Morley experiment. Einstein must have known about this experiment, if only indirectly through Lorentz' 1895 paper. However, he maintained that it had not directly affected his thinking. For him, the first-order effects were sufficient.[236] His reflections on the lack of effect of the experiment on his thinking are probably correct. The second-order effects are of crucial importance only when one is concerned with the state of motion of the aether, and by 1905 Einstein had given up the aether altogether. Still, it does seem odd that he would have known so little about *the* experiment during the period in which he was attempting to devise "a new and simple method for investigating the motion of matter relative to the light-aether"!

In his kinematical analysis, Einstein began with the concept of simultaneity. Description of the motion of a material point requires values of the coordinates as functions of time. What Einstein pointed out is that all *judgments* of time are actually of simultaneous events, in particular, of the arrival of a signal and the position of the hands on a clock. There is no problem with assuring ourselves that events in the same location are simultaneous. However, the high velocity of light has fooled us into thinking that there is no problem with distant events (i.e., because of the high velocity of light the occurrence of distant events and our receipt of the information appear instantaneous). Judgments of simultaneity at distant points are actually dependent upon the existence of synchronous clocks at distant points. Thus, we are faced with the problem of how to synchronize clocks located at distant points. We can synchronize clocks either by setting them together in the same location and then moving them away into place or by sending a signal from one to the other and back, maintaining *by definition* that the velocity of the signal is the same in both directions. With the former method we have to worry about whether transportation would disturb the synchronization. With the latter we do not have this concern, but we need to be aware of the stipulative nature of the velocity of the 'synchronizing signal', which, in Einstein's analysis, is a light signal. (It should be noted that it is unclear in his analysis what the relationship is between his definition of simultaneity and his second postulate.) Using the latter method of synchronizing the clocks, it is possible to define what is meant by 'simultaneity', and thus by 'time', within a particular reference system:

> The 'time' of an event is the reading simultaneous with the event of a clock at rest and located at the position of the event, this clock being synchronous, and indeed synchronous for all time determinations, with a specified clock at rest.[237]

It is important to see here that 'synchronous' is an 'equivalence relation': If clock A is synchronous with clock B and B with C then A is synchronous with C.

Einstein argued that we run into difficulties when we attempt to make comparisons of events described with respect to systems in motion relative to one another because we unreflectively accept that if events are simultaneous with respect to one system, they are with respect to all systems in constant relative motion to it. However, analysis of this notion of simultaneity reveals a problem. An observer located in a 'stationary' system can synchronize each of a pair of moving clocks, taken individually, with a clock located in the 'stationary' system. Applying the transitivity rule, the 'moving' clocks will be synchronous with one another, viewed with respect to the 'stationary' system. An observer in the 'moving' system can synchronize the 'moving' clocks directly, since they are stationary in her system, by taking the velocity of light to be the same for both legs of the trip. However, the *calculated* transit time forward will not equal the transit time back, if we use the classical addition of velocities theorem (Galilean transformation rules). Thus, the observer in the 'moving' system would find the two clocks not synchronous, while the observer in the 'stationary' system would find them synchronous. We have two choices at this point: Either the velocity of light is not constant for both legs of the trip in the 'moving' system or we cannot universally agree on 'synchrony' and thus on 'simultaneity' of events, and, more generally, on 'time'. Einstein chose to make the constancy of the velocity of light, for which there was a significant amount of experimental evidence, a postulate of his theory. Thus, he held that there can be no 'universal time'.

The analysis of 'simultaneity' provided the key to the formulation of the special theory of relativity. With his new conception, Einstein proceeded to derive new transformation rules for events described with respect to inertial systems. These turned out to be the same transformations, in simplified form, as Lorentz previously had derived. Einstein did not acknowledge this until later. This is not surprising given that the complete transformations for first- and second-order phenomena only appeared in Lorentz' 1904 paper, which Einstein had not read. Unlike with Lorentz' interpretation of these rules, in special relativity they are completely reciprocal. That is, when viewed from a 'stationary' system, the length of an object in motion relative to that system would appear contracted and the time indicated on a clock in relative motion to that system would be less (i.e., the clock would appear to run slow) and the same would hold true of length and time with respect to the 'stationary' system when viewed from one in motion relative to it, e.g., the system of the moving body or clock. In special relativity, the apparently paradoxical character of the situation disappears once Einstein's analysis of 'simultaneity' is applied.

Poincaré, independently in 1905, was to show reciprocity for Lorentz' version of the transformations as well.[238] However, as we have seen in the previous chapter, one crucial difference in interpretation was to remain: In relativity theory the contraction in length is not an actual physical foreshortening of bodies and there are no 'correct' lengths and times. Given Einstein's interpretation of the transformation rules, space and time no longer have absolute significance, since the values of the spatial and temporal components of an event are dependent upon the relative state of motion of the reference frame of the observer. There are no 'privileged' observers, with the correct time and coordinate values. 'Space' and 'time' are best considered as one four-dimensional 'space-time' continuum, in which events take place. Again, let me emphasize that Einstein derived these transformation rules free from any assumptions about an aether and free from any particular theory of electromagnetism or mechanics.

Einstein went on to show that Maxwell's equations are invariant with respect to the new transformation rules. Thus, a reformulation of them was not necessary – only a reinterpretation was needed. One important consequence of this reinterpretation is that what are considered separate electric and magnetic fields in Maxwell's and Lorentz' theories, should be viewed in special relativity as one electromagnetic field. What may appear as a magnetic field with respect to one reference system will appear as both an electric and magnetic field with respect to a system in motion relative to the first. Thus, it is best to view the situation in such a way that an electromagnetic field exists with respect to both systems. This interpretation solves the problem with which Einstein began the paper, namely, the asymmetrical description, required in both Maxwell's and Lorentz' theories, of the production of a current in a conducting wire when the magnetic source is moving and when the conducting wire is moving.

Einstein went on to derive a general form of aberration from the equations, as well as the known pressure of radiation and a new transverse Doppler effect. He maintained:

> All problems in the optics of moving bodies can be solved by the method here employed. The essential point is that electric and magnetic force of the light, which is influenced by a moving body, be transformed to a coordinate system at rest relative to the body.[239]

Surprisingly, he did not derive the law for the inertia of energy in this paper, but was to do so shortly afterwards in a brief paper.[240] This consequence of the special theory reinforces his rejection of the aether.[241] That is, if electromagnetic fields themselves possess inertia which increases with their energy, then they should be considered 'substance-like' themselves and not as states of substances. I will return to this shortly in the discussion of the concept of field in special relativity.

In the final section of the relativity paper, Einstein presented a new mechanical theory. Given the anisotropic nature of the new transformations, it was obvious that a reformulation of the laws of mechanics was necessary, since the relationship between force and acceleration is now dependent upon whether the force acts in the direction of the velocity of the system or in some other direction. Einstein first derived the new equations of motion for an "electron" in an electromagnetic field. He concluded that mass is not an invariant property of "electrons," but is dependent upon the state of motion of the system with respect to which it is described. He then argued:

> . . . that these results as to the mass are also valid for ponderable material points, because a ponderable material point can be made into an electron (in our sense of the word) by the addition of an *arbitrarily small* electric charge.[242]

He concluded the paper by listing a few consequences of the new equations of motion which should be experimentally detectable. The problem which remained was finding a field description of gravitation that is consistent with the new equations. This process, which was to take another ten years, will not be discussed here. Since our concern is with the electromagnetic field, the historical dimension of this book is completed with the special theory of relativity.

What is the concept of field in the special theory of relativity? First, regarding the aether, if all inertial systems are equivalent for the validity of the laws of electrodynamics, we are prevented, *in principle*, from attributing *any* state of motion (i.e., even rest) to the aether. Special relativity removes any possibility of giving a 'mechanical' or a 'dynamical' description of the aether. This goes even further than Lorentz' conception, since he felt a 'dynamical' description might be possible. Einstein later claimed that the removal of the aether concept from the description of electromagnetic processes is not strictly *required* by the special theory. It is simply "superfluous," as has been discussed. In an address, "Äther und Relativitätstheorie," given at the University of Leiden in 1920, Einstein stated what he believed the special and general theories indicate about the aether hypothesis:

> Meanwhile more careful reflection teaches that this denial of the aether is not necessarily demanded by the special principle of relativity. One can assume the existence of an aether; only one must give up attributing a definite state of motion to it; i.e., one must, through abstraction take from it the last mechanical characteristic which Lorentz had still left it.[243]

> The special principle of relativity forbids us to assume the aether as existing of particles observable through time, but the aether hypothesis itself is not in conflict with the special theory of relativity. One must only guard oneself against granting a state of motion to the aether.
>
> Certainly, from the standpoint of the special theory of relativity, the aether hypothesis appears at first to be an empty hypothesis. In the

electromagnetic field equations there appear, besides the densities of electric charge, only the field intensities. The course of electromagnetic processes in a vacuum seems to be fully determined by these internal laws, uninfluenced by other physical quantities. The electromagnetic fields appear as ultimate, irreducible realities, and for the time being it appears superfluous to postulate an homogeneous isotropic aether-medium, the fields being conceived as states of such a medium.

But on the other hand, there is a weighty argument in favor of the aether hypothesis. To deny the aether means, in the final analysis, that no physical properties whatever belong to the empty space. With this view, the fundamental facts of mechanics are not in harmony.[244]

Thus, in the special theory the aether hypothesis is "empty" since there are no physical processes associated with the vacuum itself. That is, we could keep it around, but it is even more 'ghostly' than Lorentz' aether in that it has no physical properties independent of those of the electromagnetic field. Einstein seems to have been going out of his way here to find a means of retaining the aether – perhaps out of respect for Lorentz. The "weighty argument" in favor of the aether hypothesis alluded to here comes from the general theory of relativity (pun intentional!). According to the general theory, physical processes take place in 'empty space', i.e., in space free from matter and electromagnetic fields. These processes are associated with the gravitational field and there is the possibility, given the theory, that there is no such thing as truly empty space, i.e., space free from *all* matter and fields. Here, and in other places, Einstein claimed that the name 'aether' could be associated with these processes, but the issue reduces to one of semantics. The essential point is that with the special theory the fundamental character of "the aether hypothesis" must be changed: The hypothesis of a *material* or *quasi-material* aether can no longer be supported.

The removal of the aether from the description of electromagnetic actions marks a profound change in the concept of field. It is a consequence of special relativity that no signal can propagate with a velocity greater than that of light.[245] Thus, the energy and momentum of electromagnetic actions must be in the field during the time delay in transmission, in order that these quantities be conserved.[246] It follows from this that the electromagnetic field is the seat of very complex processes, as was the case with Maxwell and Lorentz, with one crucial difference:

The electromagnetic field of special relativity is no longer a state of a material substance; rather it is an independently existing entity, which possesses energy and momentum like ponderable matter. I have exercised caution in calling the field 'substance-like' and an 'independent reality', rather than simply calling it a 'substance', because the questions of what is a 'substance' and whether in the special theory or, especially in the general theory, there is anything like the traditional notion of 'substance' are ones I would rather not discuss here.[247] They are questions that, however interesting otherwise, are beyond the scope of the present analysis. All that is required here is that we see that the field is no longer a state of some sort of matter, but is ontologically on a par with matter. In the special theory, the field is not only indispensable for the description of electromagnetic actions, as with Maxwell and Lorentz, but is also an *irreducible element of description*, in the same sense as the concept of matter in Newtonian mechanics. It is independent of matter and is free to interact with it. The electromagnetic field interacts with matter by receiving energy and momentum from it and transferring these quantities to it. In order that these quantities be conserved, it is necessary to attribute the same dynamical properties of energy/mass, momentum, and angular momentum, attributed to matter, to the electromagnetic field. As Einstein said in making a comparison between his conception and that of Lorentz:

> Only the conception of a light-aether as the bearer of electric and magnetic forces does not fit into the theory described here; that is to say, electromagnetic fields appear here not as states of some sort of matter, but rather as independently existing things, which are of the same nature as ponderable matter and have the feature of inertia in common with it.[248]

Looked at from the central problem of this book, the meaning of scientific concepts, Einstein's "only" is of course an understatement. The shift from being a "state" to being an "independently existing thing" is a major conceptual change, which, in this instance, occurred in conjunction with major changes in the fundamental concepts 'space', 'time', and, indeed, 'physical' itself. Having thus arrived at the conception of 'field' as a state of space, the historical analysis is complete.

6.3 Introduction: The gravitational field concept

Before concluding this chapter, and thus this part of the book, I will succumb, momentarily, to the urge to say something about the concept of the gravitational field as it developed in the general theory of relativity. I want to conclude this analysis with some indication of further developments in the concept of field – in the general theory, in Einstein's later thought, and in contemporary physics – in order to underscore the ongoing nature of the construction of meaning in scientific theories.

What remained for Einstein, in order to present a complete theory of relativistic mechanics, was the incorporation of gravitation into the relativistic framework. To achieve this, at least two revisions of the Newtonian equations would be necessary: (1) The equations of motion of a particle in a gravitational field would have to be generalized to the four-dimensional space-time conception of the special theory and (2) The field equation for gravitation would have to be modified since the gravitational potential in the Laplacian formulation of Newtonian mechanics responds instantaneously to changes in the density of matter and is thus not invariant under the Lorentz transformations.[249] The attempt to make these revisions led, after a considerable struggle of ten years, to the general theory of relativity.[250]

In the general theory, a new gravitational field concept was formulated, which involved even more radical changes in our conceptions of space and time. The gravitational field is an essential element of description in the general theory; it is not eliminable as is the 'potential field' of Newtonian mechanics. First, the gravitational field is not completely determined by the position of matter and other fields. Boundary conditions are needed to determine the values of the components of the 'metric tensor'.[251] By analogy with the electromagnetic field, it is quite possible (although whether or not this is actually the case is a matter of dispute) that the gravitational field can be influenced by outside radiation ('standing gravitational waves'). Thus, there is even the possibility of matter-free solutions, which do not approach the special relativistic case (components of the metric tensor are constants). There have been some claims of experimental detection of gravitational waves, such as those made by Joseph Weber at The University of Maryland, and there is a resurgence of interest in the issue today.

Most importantly, the reasons why the gravitional field is indispensable for physical description go beyond those of the electromagnetic case. The gravitational field is described solely by means of the components of the *metric* tensor. This means that the gravitational field and the structure of space-time are fused in the general theory. Even the special relativistic case must be interpreted as a special case in which the components are constant. Thus, space-time, even in this case, is not without a gravitational field; it is just a constant field. If the gravitational field were removed, i.e., the components of the metric tensor are zero, there would be no physical, topological space-time. This is quite different from the case of the electromagnetic field. Space-time without an electromagnetic field is possible. The fusion of the gravitational field and the metric leads to quite a different conception of space-time than the one formulated in the special theory. Both the structure of classical space and time and the structure of special relativistic space-time are independent of matter and fields. Both can, in principle, be conceived of as an infinite rigid body – an 'arena' in which physical events take place, but which is not affected by these events. The space-time of the general theory looses this 'rigidity' in that its structure is affected by the presence of matter and fields.

In the general theory, the gravitational field is an irreducible element of description, just as matter in Newtonian mechanics and the electromagnetic field in the special theory. The gravitational field interacts with matter in a special way: It determines the *natural* motions of matter, with 'straight-line' motion being a special case of possible natural motions. In so doing, it transmits energy and momentum to matter and receives it from matter, as is made evident by the expression for the conservation of total energy and momentum.[252] In order to satisfy the conservation laws, it becomes necessary to attribute the same dynamical properties of energy, momentum, and angular momentum, attributed to matter, to the gravitational field. In addition, as seen in the formulation of the field equations, the gravitational field itself produces gravitational effects; i.e., it has energy and thus inertial mass of its own, and so acts 'gravitatingly'. So, gravitational fields affect their own structure, which is not the case with electromagnetic fields, since they are neither charged nor magnetized and thus cannot contribute to their own source. The general theory does reveal a new feature of electromagnetic fields, though. Since they have energy, they produce gravitational effects and thus affect the structure of space-time, which, in turn, means that an electromagnetic field can affect the motion of *non-charged* matter.

These features – that it is indispensable for description and that it is an irreducible element of description – make the claim that the gravitational field is ontologically on a par with matter an extremely plausible one. It is the interpretation given to it in the present form of the theory. However, it is not the interpretation Einstein desired. His initial conception was that the gravitational field is an emanation of matter – without matter there would be no gravitational field. He quite soon came to the position that all matter should be reducible to structural properties of the space-time continuum. In fact, in his 'unified field theory', the gravitational field would have been the *only* irreducible element of description, with matter and electromagnetic fields being reduced to it.[253] That matter and fields appear ontologically on a par in the general theory, Einstein claimed to be a "dualism . . . disturbing . . . to every orderly mind."[254]

It certainly was "disturbing" to his mind! Einstein spent most of his life after the development of the general theory in an attempt to construct a 'unified field theory'. At several points he felt he had almost achieved it, at least for electromagnetism.[255] The "disturbing" nature of the situation is readily understandable: Not only is there a "dualism" of matter and fields, there is also a "dualism" of fields. Electromagnetic fields are not structural properties of the space-time continuum, but are logically independent constructions. The development of the quantum theory of matter, which is essentially a field theory of matter, and of the quantum theory of electromagnetic fields for a while seemed to indicate that perhaps the hope of total unification could, in principle, not be achieved due to irreconcilable differences in the interpretation of these fields by the theories. However, there is a resurgence of interest in the problem today and attempts at the construction of a 'grand unified field theory' make Einstein's dream again a real possibility – only, consistent with the results of this whole analysis, the 'dream' has taken on a new form true only to the 'spirit' of Einstein's dream, just as Einstein's 'unified field theory' would have been true only to the *spirit* of Faraday's dream of a 'unified force theory'.

By way of summary, let us return to Einstein's claim that the gravitational field has become the 'aether' of general relativity. Continuing his address at Leiden, he said:

> Summarizing, we can say: According to the general theory of relativity, space is endowed with physical qualities; thus, in this sense there exists an aether. According to the general theory of relativity a space without aether is unthinkable; for in such [space] there would be not only no propagation of light, but also no possibility of the existence of measuring rods and clocks, thus also no space-time intervals in the sense of physics. But this aether may not be thought of as endowed with the property that is characteristic of ponderable media, namely, to consist of particles observable through time; the concept of motion may not be applied to it.[256]

In a sense, Einstein is correct, since there is no such thing as 'empty' space with the general theory. 'Empty space', i.e, space free from matter and electromagnetic fields, is still the seat of very complex physical processes. The totality of these processes is represented by the 'impulse-energy tensor' of the general theory. We can, as Einstein suggested, use the name 'aether' for 'empty state processes', but we can also use the name 'gravitational field' or the name 'space-time'. That we can do so makes the magnitude of the conceptual change brought about by the general theory most apparent. The notion of a *material* aether no longer even makes sense. My guess is that Einstein suggested the use of the term 'aether' to symbolize the affinity, for which he always argued, between his views and the views of earlier thinkers who were convinced that electric, magnetic, and gravitational actions are not actions at a distance but are transmitted continuously through space. In light of this suggestion, I will end with Maxwell's view of what empty space is good for:

> The vast interplanetary space and interstellar regions will no longer be regarded as waste places in the universe, which the Creator has not seen fit to fill with the symbols of the manifold order of His kingdom. We shall find them to be already full of this wonderful medium; so full, that no

human power can remove it from the smallest portion of space, or produce the slightest flaw in its infinite continuity. It extends unbroken from star to star; and when a molecule of hydrogen vibrates in the dog star, the medium receives the impulse of these vibrations; and after carrying them in its immense bosom for three years, delivers them in due course, regular order and full tale into the spectroscope of Mr. Huggings at Tulse Hill.[257]

. . . or perhaps into the apparatus of Mr. Weber at College Park?

PART III

The Making of Meaning: A Proposal

Words are not just wind.
Words have something to say.
But if what they have to say
is not fixed, then do they really
say something? Or do they say
nothing?

Chuang Tzu

Introduction

Having surveyed the most influential philosophical accounts of the nature of meaning in scientific theories and discussed their deficiencies, and having analyzed the major episodes in the history of the construction of the electromagnetic field concept, we need now to see if it is possible to construct an adequate 'working hypothesis' for how to conceive of 'meaning' and 'meaning change' in scientific theories. The key to the development of an adequate conception lies in the recognition, amply underscored by the historical analysis, that scientific concepts do not arise in some 'creative leap'; rather, they are constructed from a changing network of beliefs and problems. In the conception to be presented here, the actual process by which meaning is construed in science forms a significant part of the notions of 'meaning' and 'meaning change'. Thus, I will begin by discussing selected features of the construction of the current electromagnetic field concept and go on to make a proposal for how to conceive of meaning in scientific theories.

CHAPTER 7

Meaning in scientific practice

7.1 Constructing the electromagnetic field concept

Our historical analysis shows that the construction of the present concept of electromagnetic field took place in three periods, or phases, which I will call: (1) 'heuristic guide', (2) 'elaborational', and (3) 'philosophical'. The contributions of Faraday and of Maxwell, until his third paper, come under phase (1); those of Maxwell, beginning with the third paper, and of Lorentz, under phase (2); and those of Einstein fall under phase (3). The reasons for these groupings and for the choice of name for each phase will become apparent as we discuss the relevant details that have emerged from the historical analysis.

In the first phase, a new way of conceptualizing electric and magnetic actions was introduced and taken to the point where it became a scientifically viable alternative to the action-at-a-distance conception of these forces. Faraday's initial conception was a rather vague, general, and intuitive speculation concerning the possible location of the action in the surrounding space; i.e., that some processes in, or some state of, space may be necessary to the description of such actions. This initial conception emerged out of the combined network of beliefs and problems we have discussed. Central among these are: Faraday's prior reflections on the nature of matter and force; his belief in the primacy of experiment; and the problem of how to interpret the phenomenon of "electromagnetic rotations". From this background he hypothesized that (1) the observed rotational motion in the Oersted discovery is simple and (2) the motion results from activity in the space around the needle and the current-carrying wire. These hypotheses led to his attempts to articulate what could be meant by 'activity in space' and to find experimental confirmation of it.

What set Faraday apart from most of those working on the problem of interpreting the new discoveries in electricity and magnetism was his recognition that, indeed, as his own interpretation was speculative, the conception of forces which lay behind the complex description of the action in terms of attractive and repulsive forces was no less so: Along with Newton he held that the action-at-a-distance conception of forces is a speculation and not a very satisfactory one; therefore, he did not attempt to 'fit' this and other phenomena into the Newtonian framework. Throughout his three periods of research he endeavored to devise experiments and to construct arguments *for* the new conception and *against* the action-at-a-distance conception. In the process he developed what he meant by 'continuous, progressive transmission of the action', with the central question being that of how or by what processes such actions could be possible. He articulated this notion with the help of a concrete visual image and of vague analogies between electric and magnetic actions and 'known' progressive phenomena.

Before discussing his response, though, it should be pointed out that, as the reader is already aware from Chapter 3, Faraday did not use the term 'field' himself, except to define it in terms of the 'lines of force' late in his *Researches.* This, however, does not present a problem for us since one can have a concept and not use the term which customarily represents it.[1] The term 'field' came into use, in this context, through Thomson's discussion of Faraday's work.[2] Most likely, Faraday preferred to use the expression 'lines of force' because of the influence of the visual image on his thinking.

The image for representing the action came with Faraday's discovery of electromagnetic induction. The initial problem was how to account for the fact that a change in the magnetic force produces a current, whereas a constant force does not (since a constant current does produce a magnetic force). His first attempt to solve this problem was to introduce the notion of an "electrotonic state," but the state could not be detected experimentally. The image came while addressing a second problem: how a current arises in a conductor simply through the motion of the conductor or the magnetic source. His answer was that induction takes place by the "cutting" of the lines of force, either by the motion of the conductor or of the magnetic source, or by the *motion of the lines* across a conductor.

The image acted both *heuristically* – as a guide to his experimentation and reasoning – and *descriptively* – as an account of the actual mode of transmission of the actions. The tension between these two, quite different roles of the image remained throughout his work, since Faraday was not able to establish the existence of the lines in empty space. A curious thing about the image is that it is discrete, whereas the conception it represents is continuous. The discreteness of the image influenced Faraday's experimentation, his arguments, his choice of terminology, and his choice of analogy.

We have not paid much attention to the details of the specific experiments conducted by Faraday; rather, we have focused on the results and the conclusions he drew from these. That is, our emphasis has been on the conceptual/theoretical dimension of Faraday's work. A more complete analysis would include the experimental/procedural dimension, and would show the influence of the image in his experimental design.[3] Some of this can be seen from the arguments we have considered. Let me remind the reader of the most important. Faraday argued that there is a quantitative relationship between the number of lines cut and the intensity of the induced force. The discreteness of the image was perhaps responsible for the error in his formulation of that relationship.[4] A central and repeated argument Faraday gave against the actions being distance actions is that action at a distance is supposed to take place in straight lines and to be unaffected by any intervening medium, while electric and magnetic actions seem to be in curved lines and are affected by the nature of the medium. He also argued that a connection between static and current electricity might be found in the expanding and collapsing of the lines and that conceiving of the lines of magnetic force as being 'conducted' with differing ease through various types of magnetic media could provide a unified conception of magnetic actions. He speculated that light, and possibly gravitation, could be produced by a shaking of the lines.

His choice of terminology also reflects the influence of the discrete image, e.g., 'number cut', 'expanding lines', 'contracting lines', 'shaking', 'vibrating', 'turning corners', etc. Indeed, some of his discussions indicate that he looked at the lines as individual 'strings', although he never said this explicitly. The analogies he used most frequently were with other 'line-like' phenomena, such as rings formed after dropping a stone in water, vibrations of sound in air, conduction through a wire, radiant heat, and light rays. Faraday used these analogies as aids in expressing what he meant by continuous, progressive action in or by means of lines of force and to help him articulate what could be meant by such action. In this 'articulating' role, the analogies are used heuristically; i.e., they provide a temporary way of conceptualizing the unknown actions. On one level, the types of actions are identical: progressive and continuous transmission from source to receiver (Faraday's assumption); on another, the lines are only similar to rings, waves, rays, wires, etc. Thus, we see these analogies functioning in two ways: on a more concrete level, they provide a means of visualizing and describing the new conception of electric and magnetic actions in terms of familiar actions; and on a more abstract level, both the new and the familiar actions are assumed to be of the same nature, i.e., continuous and progressive. It is at the concrete level that we see a little-discussed function of analogies in science: Analogies are used to give temporary meaning to a vague, unarticulated conception, and they are also used to assist in the construction of its meaning.[5] This use of analogy is more evident in Maxwell's contribution, as we have seen and will discuss further.

Finally, Faraday's full field conception is that of a 'unified field of force', in which all the forces of nature are transmitted through lines of force. Material particles are point centers of converging lines of force, with all forces conveyed through them, and with every 'particle' thus connected with every other, its sphere of influence being "the whole solar system." I have argued that this conception coalesced during his second period of research, in his attempt to formulate a conception of the continuous transmission of force in dielectric media: Electrostatic action is through "contiguous particles," but these "particles" are really connected by their lines of force. In his final period of research, during which he extended this conception to magnetic induction, he felt there was some evidence, albeit not 'crucial' evidence, for the existence of the lines in space free from any material medium. As we have seen, he preferred not to include an aether in this space since his experimental findings did not provide evidence for it and his speculation did not require it, i.e., if the lines of force are themselves substances, there is no need for a substratum through which they transmit actions.

Faraday's conception, however, was not to be carried on in its entirety by those who came after him in electromagnetism. Given its non-mathematical form, the full conception was not scientifically viable. Whether or not it could have been (or could be) formulated mathematically, it was not in fact because Maxwell did not fully "grasp" its details. This, in turn, was so because Maxwell, although profoundly influenced by the general form of Faraday's conception, approached the problem of how to formulate continuous, progressive transmission of electric and magnetic actions mathematically from a rather different network of beliefs and problems. Central in Maxwell's network are: the Cambridge mathematical tradition of continuum mechanics; Thomson's use of "physical analogy" in his attempts to interpret Faraday's speculations; the difficulties with the action-at-a-distance conception as formulated by Weber; the appeal of Faraday's continuous-action conception to the "physical reasoner"; and the problem of devising a mathematical formulation of the general conception which would include the crucial time delay in transmission.

The lines of force did not have a strong effect on Maxwell's thought, since his problem, after the first paper, was that of providing an account of the stresses and strains thought to be associated with electric and magnetic actions, i.e., his concern was more with what Faraday had called the "electro-tonic state" than with the lines themselves. For Maxwell the latter were just the visible effects of the underlying stresses and strains in a mechanical medium – a Newtonian aether. In 'continuum mechanics' satisfactory mathematical ways had been created for representing other continuous, progressive transmissions of force in media, such as fluids, and elastic solids, even though little was known about their underlying structure (e.g., molecular or not). Thus, the obvious approach for Maxwell was to treat the aether as such a mechanical system and to conceive of electric and magnetic fields as states of it.

As was discussed in detail in Chapter 4, Maxwell made use of several physical analogies, first to give a representation of the configuration of the forces in the medium, and then to provide a formal account of how those forces could be produced. He used his analogies *heuristically*: to provide temporary physical and mathematical meaning to the general conception of stresses and strains in space free from ordinary matter, and to assist in the articulation of what could be meant by 'continuous, progressive transmission of electric and magnetic forces through a mechanical aether'. We can see that his analogies also function on two levels. At an abstract level, the types of force are assumed to be identical, i.e., electric and magnetic forces are assumed to be conveyed in a continuous, progressive manner as with forces in a fluid, or with forces in an elastic medium under stress. On the other hand, at the concrete level, the phenomena are only similar, e.g., the assumed stresses and strains of electromagnetism are similar in nature to those produced in an elastic medium subject to the constraints imposed by Maxwell. Identity at the abstract level is needed for the choice of analogy; this part of the meaning is shared by the conception under construction and the conceptions borrowed from another domain. It is primarily the similarities at the concrete level which provide assistance in the construction of the meaning of the new concept. At this level the analogies perform what might be called an 'articulating function': They provide temporary meaning until the new meaning has been sufficiently articulated and also provide a basis from which to construct the new meaning. In the case at hand, the analogies acted as a guide to reasoning about the problem of how to conceive of electric and magnetic forces in a medium until the field equations were formulated. Once a field conception of these forces had been elaborated to a scientifically adequate extent, the analogies were no longer needed in their 'articulating' capacity, i.e., what Maxwell called "physical analogies" were no longer needed. Of course analogies continue to be of use in providing an intuitive understanding of well-articulated concepts, but this is a different function and these are not "physical analogies." We can say, for example, that the transmission of energy in an electromagnetic field is like flow in a fluid, but there the analogy performs an explanatory function rather than an articulating one.

As we have seen, the use of "physical analogies" led Maxwell to construct a scientifically viable concept of electromagnetic field: Electric and magnetic forces are propagated with the velocity of light in a Newtonian aether according to a specific set of equations. (It is 'scientifically viable' in that it is formulated mathematically and has empirically testable consequences, and so provides a real alternative to the action-at-a-distance conception.) These field equations, however, give no indication of what the underlying 'mechanism' may be like. This fact had two important consequences. On the one hand, it led many of those who believed in the continuous-action conception to search for such a mechanism and led finally to the conclusion that there is none; while, on the other hand, this same fact also made it possible to enter the second 'phase' of development without having to answer the question concerning the 'mechanism'. Phase (2) began with Maxwell's third paper in which, as I have argued, the transition was made from a 'mechanical' analysis to a 'dynamical' one by treating the aether as a "connected system." So, with the mathematical formulation a dimension was added to the meaning of 'electromagnetic field' that made it unnecessary to find out exactly what kinds of processes were involved before proceeding to the elaboration, clarification, and extension of the new concept (just as had been the case with Newton's concept of gravitation). Maxwell's 'dynamical' analysis masked the non-Newtonian nature of the system, which came to the fore in Lorentz' analysis. Despite its mathematical precision, Maxwell's field concept is still vague, not only with respect to the nature of the underlying processes, but also regarding the differences between aether and 'ordinary' matter and regarding the nature of charge and its relation to the field. It was Lorentz' endeavor to clarify these – from his network of beliefs and problems – which led to the derivation of the non-Newtonian 'Lorentz force' and to the subsequent non-Newtonian features of his "theory of electrons."

As has been said, the problem of constructing an account of the mechanical processes within the aether which could produce the state now known as the 'electromagnetic field' was central in late-19th and early-20th century British science. With the electromagnetic theory of light, the problem of the nature of these processes became coupled with problems concerning the luminiferous aether; in particular, with the problem concerning the global and local state of motion of the aether. The next significant change in meaning came about largely as the result of Lorentz' attempt to solve mathematical problems posed by mathematical and physical constraints on his conception of the aether as immobile (locally and globally). Lorentz brought elements of both the action-at-a-distance conception and the continuous-action conception of the transmission of electric and magnetic actions to his first problem: to account for reflection and refraction of light, given the electromagnetic field equations. He used Helmholtz' reformulation of Maxwell's analysis, i.e., his action-at-a-distance aether, since he found it more comprehensible than Maxwell's formulation, but later accepted the continuous, progressive transmission conception for reasons we have discussed. The results of this first analysis, especially the conclusion of the superiority of the electromagnetic theory of light over the elastic solid theory, made it imperative that he elaborate upon questions left unanswered or only partially addressed by Maxwell: the state of motion of the aether (since electromagnetic fields in a stationary aether would differ from those in a moving one), the distinction between aether and other dielectrics, and the nature of charged matter and its relation to the aether.

The chief factors in Lorentz' network of beliefs and problems are: his 'instrumental' attitude towards, and his 'constructive' approach to, physics; his focus on problems in optics; his assumption of Maxwell's equations as fundamental; the problem of separating the aether from 'ordinary' matter and the nature of aether – matter interactions; his assumption of the existence of 'molecules' (e.g., kinetic theory of gases); and his assessment that a Fresnel-like aether is superior to that of Stokes. Lorentz tried to combine a conception of the continuous and progressive transmission of electromagnetic actions with a conception of matter as systems of charged particles. He assumed that the aether is rigid, which required that he disregard an important part of Maxwell's field concept: There could be no mechanical description of the motion of the aether 'particles'. The assumption of a rigid aether led to the local violation of action and reaction in aether – "ion" interactions, which makes the field a state of a non-Newtonian aether. It was this conception of the aether – as one whose only mechanical property is 'absolute' rest – which formed part of the background network from which Einstein was able to remove 'aether' completely from the description of the transmission of electric and magnetic actions, while retaining continuous, progressive conveyance of electromagnetic forces.

As has been discussed, there were significant problems and questions of interpretation created by Lorentz' *rapprochement.* Given developments in other areas of physics as well, it became essential to understand how the new field conception of electric and magnetic forces fits into the rest of physics: If matter is composed of charged particles, then there needs to be a unified framework for mechanics and electrodynamics, and if we accept the kinetic theory of gases, then thermodynamics must also be incorporated into such a framework. Lorentz did not undertake the construction of a unified theory, although he was master of all areas of physics; other theorists tried to devise a comprehensive electromagnetic theory in which mechanical forces would be reduced to states of the electromagnetic aether, while others, such as Poincaré, still sought a unified Newtonian theory. It was Einstein's ten-year critical reflection on the foundations of all of physics which led him to see the inconsistencies and which led, in the end, to the present conception of field. In this third 'phase' of development, the foundations upon which the new field conception of forces had been constructed were reassessed in order to bring it into the whole of physical theory.

The main elements in the background of beliefs and problems which led to Einstein's elimination of the aether are: how to account for the Aarau *Gedanken* experiment; the philosophies of Hume and Mach; the problem concerning the state of motion of the aether (his possible experiments); Lorentz' rigid aether and the "disturbing dualism" of particle and field in his theory; his (Einstein's) realization of the repercussions of Planck's quantization of the energy of radiation for both mechanics and electromagnetism; and his work on Brownian motion and the photoelectric effect. Einstein's critical reflections directed him to the conclusion that the aether is unnecessary as a physical concept, i.e., that the electromagnetic field can be conceived of as a state of space itself. Given his analysis of space and time, the aether cannot function as an 'absolute' reference frame and, given his derivation of the 'Lorentz transformations' from the analysis of space and time (rather than from specific electrodynamical problems as was the case with Lorentz' derivation), the aether is not needed in its causal role, in particular, for explaining the contraction in length.

With Einstein's concept of field, the aether is "superfluous"; however, although we have come once again to a concept of electromagnetic field without the notion of an aether, we have not returned to Faraday's field concept. Indeed, we began our analysis of the construction of the meaning of 'field' with the tacit presumption that there is *a* field concept whose history we were to examine, but now we must conclude that our present concept is just that – the *present* concept and not *the* concept. Yet we have also shown a definite relationship between the present concept (Einstein's) and the field concepts of Faraday, Maxwell, and Lorentz. So, in the language of Part I, we have 'meaning variance' and 'commensurability'. We need now to return to our initial problem: how to formulate the notions of 'meaning' and 'meaning change' in a way commensurate with scientific practices.

7.2 'Meaning schemata' and commensurability

In concluding Part I, I argued that a conception of meaning adequate for scientific theories could not be created merely from a study of the nature and necessities of language, but only in conjunction with an analysis of actual scientific practices concerning meaning. As we have seen, these practices are such that it is not possible to separate questions of meaning from the network of beliefs (theoretical, methodological, metaphysical, common sense) and problems (theoretical, experimental, and metaphysical) which provides the 'motive force' in meaning construction. Each of the field concepts considered here must be viewed within the context of its network. In the summary of the construction of the present concept of field in Section 7.1, we reiterated the key features of each of the networks; let me now focus on the results; i.e., the meaning of each electromagnetic field concept considered.

Faraday's electromagnetic field ("physical lines of force") is primarily a qualitative concept, i.e., of unknown mathematical structure. The field is responsible for the continuous, progressive transmission of electric and magnetic forces through space. It is either a property (of an aether unlike 'ordinary' matter) or a substance (state of "mere space"), with Faraday's preference being to consider the lines of force as substances and force as the only substance. Thus, in his most speculative conception, the field concept would become a broader concept comprising all the forces of nature. For Maxwell, the electromagnetic field conveys electric and magnetic forces and light according to a specific set of mathematical formulae ('Maxwell's equations'). It is a state of a mechanical aether (not clearly differentiated from 'ordinary' matter), but the mechanical processes of which it is the manifestation are unknown stresses and strains in the medium. With Lorentz, the electromagnetic field is a state of a rigid (non-mechanical) aether and is produced in the interaction of charged matter ("ions") and the aether. This interaction is governed by Maxwell's equations and the non-Newtonian 'Lorentz force'. Transformations between electrodynamic and electrostatic systems follow non-Galilean rules. In Einstein's special theory of relativity, the electromagnetic field has the mathematical structure of Lorentz' field concept, but it is not a state of an aether, rather it is 'ontologically on a par' with matter. As has been indicated, Einstein went on to construct a gravitational field concept and his full field conception was that of a 'unified field' in which the electromagnetic field and matter would also represent the structure of space-time. We ended our 'story' with Einstein's concept because it is the present concept of electromagnetic field; however, given what appears to be the open-ended nature of meaning construction in science, it is possible that at some future point our concept will seem as archaic as the 'aether-field' seems now.

We must return, now, to the question posed at the end of the previous section: What justifies our calling all of these conceptions 'electromagnetic field' and claiming that they are related, i.e., that there is a traceable line of descent from Faraday's field concept to that of Einstein? An immediate, simple, and quite correct response is that common to all the 'phases' considered here is the central problem of how to conceive of the continuous and progressive transmission of electric and magnetic forces through space: Faraday, Maxwell, Lorentz, and Einstein were all reasoning about the same basic problem. I want to go further, though, and propose a general conception of 'meaning' and 'meaning change' for scientific theories that will provide an elaboration of the 'simple' answer to the question, while at the same time providing an alternative to current philosophical conceptions of meaning in scientific theories. Of course, one cannot and should not make strong claims on the basis of one study – no matter how comprehensive and significant the concept studied – but one should also not be too hesitant, for it is possible to combine the historical data with considerations about meaning taken from other fields and, on the basis of these together, we can devise a proposal – a 'working hypothesis' – which can be assessed and refined in the light of more data.

At the outset we need to realize that behind the philosophical conceptions of meaning we have surveyed in Part I is a particular notion of 'concept' (or 'meaning of a term' for those who do not like to speak of 'concepts'): the notion that a concept is a set of necessary and sufficient defining conditions. This notion does not accord well with our data for the field concept. We could say, e.g., that in each case there is a different set of necessary and sufficient conditions for 'electromagnetic field' (assuming that we could sift them out of each of the networks), but this would not account for the developmental process we have discerned. Indeed, with four separate definitions, we would be led right back to incommensurability. If, on the other hand, we would say that the necessary and sufficient conditions are those of Einstein's concept, so there is really no 'field' before then, this would account neither for the fact that the others were considered and functioned, in practice, as electromagnetic field concepts nor for the 'open-endedness' of concept formation, i.e., we could have, and still could develop, quite a different concept and yet consider it an electromagnetic field concept. So, this particular notion of concept does not work well. However, how to conceive of concepts is not the philosopher's problem alone. There is a rather extensive debate in the literature of contemporary cognitive psychology over just what a concept is that can assist us here.[6] The positions taken are: (a) the 'classical' view that a concept is a set of necessary and sufficient defining conditions; (b) the 'prototype' or 'probabilistic' view that a concept is a set of family resemblances, with instances of the concept varying in the degree to which they share certain properties; and (c) the 'exemplar' view that a concept is represented by its paradigmatic cases. I propose that, as the first ingredient in the new conception of meaning, we adopt, provisionally, a prototype view of concepts since it accords best with our data. Although this view of a concept is not without problems, they seem not as intractable as the widely recognized problems regarding the specification of necessary and sufficient conditions for concepts in general (i.e., those not 'explicitly defined') and those regarding how to choose 'paradigmatic cases' and how to specify the relationship between the supposed exemplars of a concept (i.e., they form a set of disjunctions).[7]

With this view of a concept in hand, we need next to consider the descriptive/explanatory function of concepts in scientific theories. For example, briefly put, the electromagnetic field concept provides a description of the transmission of electric and magnetic forces and an explanation for how such continuous, progressive action is possible. Now just what constitutes a 'description' and an 'explanation' has been discussed for centuries in the philosophical literature. In the context of our study, however, we may bypass the subtleties of these notions and focus on the relationship between this descriptive/explanatory function of concepts and the nature of meaning in science. For us, it is sufficient to say that as descriptive/explanatory systems they need to provide answers for such questions as: What does it do? How does it do it? What is its function? What effects does it produce? What kind of 'stuff' is it? How can it be located? etc.

We are now in a position to make the following 'working hypothesis': The meaning of a scientific concept is a two-dimensional array which is constructed on the basis of its descriptive/explanatory function as it develops over time. I will call this array a 'meaning schema'. A 'meaning schema' for a particular concept, would contain, width-wise, a summary of the features of each instance and, length-wise, a summary of the changes over time. To construct such an array we must decide what features should be considered 'part of the meaning', which for us will be found by considering the function of concepts in science and not by considering what is 'essential' and 'non-essential' to the meaning. I suggest we make a first selection for these features by adopting Aristotle's criteria for a good explanation, i.e., that it should include the following factors: 'stuff', 'function', 'structure', and 'causal power' (without the Aristotelian metaphysical implications), since it accords well with the questions for which concepts, in their descriptive/explanatory role, should provide answers.[8] Here, 'stuff' includes what it is (with ontological status and reference); 'function' includes what it does; 'structure' includes mathematical structure; and 'causal power' includes its effects. Of course, not all of these features will be specifiable for each concept in each phase of its construction. Table 3 provides a sketch of the 'meaning schema' for the phases of 'electromagnetic field' we have considered.

Accepting not necessarily the specific form provided but at least the 'spirit' of the notion of a 'meaning schema', we can give the question concerning the relationship between the 'different' field concepts a more concrete form: What is the connection between earlier and later forms of a concept as represented by its 'meaning schema'? For an answer to this I will borrow a notion of a 'chain-of-reasoning connection' ('*COR*') from Shapere.[9] Shapere has claimed for some time that the continuity of meaning for scientific terms is to be found in the reasons for the changes that were made. This claim, however, has not been taken up by those addressing the problem of incommensurability of meaning. My conception of meaning is a development from his claim (though one with which he might not concur). For each 'meaning schema', its earlier and later forms are both determined by and connected through the reasoning for the initial introduction and subsequent alterations of the concept. As we have seen, this reasoning takes place within a changing network of beliefs and problems. Although I claimed at the outset that concrete histories of scientific practices concerning meaning are essential to forming an adequate conception of meaning in science, we can now see that such histories are an essential part of that conception as well: Analyses of the actual construction of scientific concepts reveal the 'chains-of-reasoning' in their construction.

Table 3. 'Electromagnetic field'.

Stuff	Function	Structure	Causal power
force in region of space (lines are substances? states of 'aether'?)	transmits electric and magnetic actions (light? gravity?)	unknown	certain electric and magnetic effects
C *O* *R*	*C* *O* *R*	*C* *O* *R*	*C* *O* *R*
mechanical processes in quasi-material medium (aether)	transmits electric and magnetic actions (now including light)	Maxwell's equations	all electric and magnetic effects, optical effects, radiant heat, etc.
C *O* *R*	*C* *O* *R*	*C* *O* *R*	*C* *O* *R*
state of immobile aether (non-mechanical)	same	same plus Lorentz force	same
C *O* *R*	*C* *O* *R*	*C* *O* *R*	*C* *O* *R*
state of space	same	same but relativistic interpretation	same

C
O = 'chain-of-reasoning connection'.
R

The conception of scientific concepts as 'meaning schemata' with earlier and later forms connected by 'chains-of-reasoning' provides a viable philosophical counterpart to the response of many scientists and historians to the philosopher's problem of incommensurability of meaning: There is meaning variance, but there is also a significant degree of commensurability. The cost to philosophers is the revision of ideas of what is and is not relevant to the philosophical enterprise. In order to do justice to science the idea that meanings are independent of the processes through which they are constructed must be abandoned.

In sketching the notion of a 'meaning schema' we have done what we set out to do in this book: to propose a conception of meaning in accord with scientific practice. How adequate this conception is must be judged by how useful it is in helping us understand and account for meaning in scientific theories.

Notes

Chapter 1

1. For an extensive discussion of the 'standard account', see Suppe (1977).
2. See, e.g., Quine (1969) and Quine (1973).
3. Lewis (1952, 1956) and Carnap (1956a and b).
4. Lewis (1956).
5. Carnap (1956a), p. 213.

Chapter 2

6. Duhem (1962), esp. pp. 184–7.
7. Quine (1973), p. 38.
8. Quine (1969), p. 89.
9. Kuhn (1962) and Feyerabend (1962).
10. For a more detailed discussion of this attempt, see my (1979).
11. Quine (1969), p. 89.
12. Nersessian (1979).
13. Feyerabend (1956b), p. 242 and (1965a), p. 203.
14. Feyerabend (1962), p. 36 and (1965a), pp. 118–212.
15. Feyerabend has offered some suggestions for ways of 'evading' incommensurability, but at worst they are incoherent and at best they are inconsistent with his own arguments. He, himself, does not take them seriously, but rather, believes we will simply have to live with the problem. So, we need not consider them here.
16. For a detailed discussion, see Shapere (1966).
17. There is a third 'response', i.e., the 'non-statement' view of theories of Sneed (1971) and Stegmüller (1975 and 1976). Sneed did not develop his conception as a response to the incommensurability problem; it was Stegmüller who saw it as providing a way around the problem. They are more concerned with analyzing structural relations than with meaning. Their treatment of semantics is not incompatible with my approach – we are concerned with different levels. Their conception

does not account for the developmental aspect of meaning, which is my major concern.

18. Scheffler (1967).
19. Kripke (1972) and Putnam (1973, 1975).
20. The term 'natural kind' is used to designate naturally existing, observational entities, such as water.
21. Some of this analysis is based on an unpublished paper written with H. Philipse.
22. M. Fehér argues, in an unpublished paper "Meaning variance and scientific realism," that there is a serious problem of unwarranted ontological assumptions with all versions of the causal theory.
23. Putnam (1979), p. 233.
24. See Shapere (1982a). I agree with his argument, in principle, although I would argue that his 'paradox' arises because of an overemphasis on the synchronic aspect of the causal theory.
25. Ibid.
26. There have been a few recent analyses along the lines I propose. See, e.g., Shapere's analysis of the notion of observation (1982b).
27. Shapere (1981), p. 1.

Chapter 3

1. For an interesting discussion of scientific problems and their relationship to metaphysical beliefs, see Agassi (1975a).
2. See, e.g., Faraday (1831), 2429, 2430. (Unless preceded by 'p' all references to the *Researches* are given by paragraph number.)
3. It should be noted that there is a fundamental disagreement in the literature as to when Faraday 'had' his field concept (see Agassi (1971), Berkson (1974), Gooding (1980b, 1981), and Williams (1964, 1975)). It seems to me that the real source of the disagreement is the lack of an account of what it means to 'have' a concept. My own position is that 'having' a concept is a matter of degree, in which case most of the dispute dissolves.
4. Faraday (1821a).
5. Bence-Jones (1870), 1, p. 355.
6. Agassi (1971), Berkson (1974), Gooding (1980b), Heimann (1971), Levere (1968), Spencer (1967), Williams (1964).
7. Portions of these lectures are published in Bence-Jones (1870). The manuscripts for Faraday's chemical lectures are at the Society for Electrical Engineers in London.
8. Faraday probably became familiar with Boscovich's views through Davy.
9. Levere (1968).
10. "A speculation touching electric conduction and the nature of matter," (1844), in Faraday (1831), 2, pp. 284–293.

11. Thus, Faraday's position differs from that of Boscovich in that the latter held that atoms are point centers surrounded by attractive and repulsive forces.
12. Bence-Jones (1870), 1, p. 213.
13. Op cit., note 10.
14. My position is in disagreement with Agassi (1971) and Berkson (1974) who argue that Faraday worked out this conception, fully, at the time of his discovery of electromagnetic induction. I also disagree with Gooding (1978, 1981) who argues that Faraday came to this position only in the final period of research, in connection with his study of magnetic induction. Williams (1964) is unclear as to when Faraday had it. My own position, which will be argued for here, is that significant elements of the conception were present in his early research, e.g., the possibility of a new description of the actions and the importance of the lines. Most of the conception coalesced in his research into electrostatic induction, in particular, his speculations concerning the nature of particles and the unity of forces. The 'final' conception came with his research into magnetic induction, which is the period in which he attempted to clarify his views and in which he allowed himself to be more openly speculative.
15. "On some new electro-magnetical motions, and on the theory of magnetism," (1821), in Faraday (1831), 2, pp. 127–151.
16. Bence-Jones (1870), 1, p. 356.
17. Reproduced in Williams (1964), p. 181.
18. Faraday (1831), 60.
19. Ibid., 70.
20. Ibid., 217.
21. The relationship Faraday formulated is not quite right. Looked at mathematically, it is only in very rare circumstances that the lines of force are closed curves as Faraday assumed. In the usual case, the lines of force spiral infinitely in a finite volume. If we look at a finite surface, each line comes arbitrarily close to every other point on the surface, i.e., if a line crosses the surface once, it crosses it infinitely many times. Thus, because of the 'spiraling', as opposed to 'closedness', the same line may be cut repeatedly. (This would be so even if the lines were closed because the line might wind about before it closed on itself.) To remedy the defect, we should have to speak of 'the number of cuttings' rather than 'the number of lines cut'. Another defect in this concept used as a quantitative measure is that 'number of lines' is an integer while 'field intensity' is a continuous function. We are able only to determine the approximate intensity of the force in terms of lines of force. This problem can be dealt with by replacing 'number of' with a discrete measure, which is essentially what Maxwell showed how to do in "On Faraday's Lines of Force."
22. Faraday (1831), 220.
23. Faraday (1832), January 21 or 22, 1823.

24. Faraday (1831), 238.
25. Ibid., 524.
26. Ibid., 1164, 1165.
27. Ibid., 1320, 1368–69. The clearest picture is presented in his 1846 "Thoughts on ray-vibrations," in (1831), 3, pp. 447–452. See also Gooding's interesting discussion of this (1980b).
28. Faraday (1831), 1658.
29. "A letter to Prof. Faraday, on certain theoretical opinions." R. Hare in Faraday (1831), 2, pp. 251–261; Berkson (1974), Gooding (1978), Levere (1968), Spencer (1967).
30. Faraday (1831), 1615, 1616, 1665.
31. Op cit., note 29.
32. Faraday (1831), 2, p. 262.
33. Ibid., p. 267.
34. Ibid., 1231.
35. Faraday, himself, did feel that electrical brush discharge indicates the *prior* existence of the lines of force. Ibid., 1449.
36. Ibid., 1326. See also 1320.
37. For a detailed analysis see Gooding (1980b). See also Agassi (1971), p. 36 and Williams (1964), p. 181.
38. See Faraday (1831), 1362–1370.
39. Ibid., 1658.
40. Op cit., note 27.
41. Gooding (1978).
42. Op cit., note 10.
43. Ibid., p. 294. Further support for this move is provided by the fact that Faraday did not see any inconsistency in his reply to Hare. Given this conception of particles there is no inconsistency.
44. Bence-Jones (1870), p. 177.
45. Faraday attributed his position to Boscovich, but, as we have discussed, there are important differences. For detailed discussions see Spencer (1967) and Williams (1964).
46. See "Thoughts . . . ," op cit., note 27 and "On the conservation of force" (1857) in Faraday (1859).
47. Faraday (1831), 2146, 2221.
48. Ibid., p. 2 footnote.
49. Ibid., 2429, 2430.
50. Ibid., 2797.
51. Ibid., 3074.
52. Ibid., 3175.
53. We should note, though, that the lines of force are neither transmitters of force along themselves nor paths of transmission of force at all in Maxwell's theory. The electromagnetic field is not propagated along its lines of force, but orthogonal to both **E** and **B**. Faraday seems to have formulated some vague analogy between lines of force and light rays, hence his confusion between 'curved lines of force'

and transmission of action (either 'vehicles' or 'paths') along curved lines.

54. Faraday (1831), 3, pp. 407–437.
55. Ibid., 3258.
56. Ibid., 3246.
57. Ibid., 3251.
58. Ibid., 3269.
59. Ibid., 3254.
60. Ibid., 3258.
61. Ibid., 3263.
62. Ibid., 3277.
63. Ibid., 3278.
64. November 29, 1839 to R. Phillips in Williams et al. (1971), p. 211.
65. Faraday (1831), 3, p. 528.
66. Ibid., 3302.
67. He argued, e.g., in his "Conservation of Forces," that the "[m]athematical mind has no advantage in perceiving new principles of action." Op cit., note 46.
68. See, e.g., what Faraday has to say about the light aether at the beginning of Faraday (1831), 3.
69. Ibid., 3, p. 443. He also speculates in his "Thoughts . . ." that the aether is nothing more than forces; but he did not commit himself to this view, so neither can we.
70. Ibid., 3, 3244.

Chapter 4

71. Campbell and Garnett (1969), p. xxii.
72. Larmor (1937). See letters of November 13, 1854; May 15, 1855, p. 8 and p. 11; and September 13, 1855, p. 17.
73. Ibid., September 13, 1855, pp. 17–18. It should also be pointed out that although Maxwell did not discuss it, he most likely started with the assumption of a Newtonian aether because of the influence of Thomson and others at Cambridge. By the time Maxwell began his work there was an established community of mathematical physicists there, whose major concerns were with continuous media (continuum mechanics), such as heat and elastic solids. They had incorporated the light aether into the framework of continuum mechanics. Thomson, and subsequently Maxwell, extended notions developed in that framework to the electromagnetic aether.
74. That Maxwell's analogy is a particularly good one in the electric and magnetic cases can be seen by the following. That a fluid is incompressible means that the density 'ρ' is everywhere the same, i.e., $\rho = \text{const}$. According to the equation of continuity, $\text{div}\, \rho\mathbf{v} + \partial\rho/\partial t = 0$, where $\mathbf{v}$ is the velocity field of the fluid. Since $\rho = \text{const.}$, what this reduces to is

div $\mathbf{v} = 0$. By substituting the magnetic intensity $\mathbf{B}$ for $\mathbf{v}$, we get div $\mathbf{B} = 0$, which is true. By substituting the electric intensity $\mathbf{E}$ for $\mathbf{v}$ we get div $\mathbf{E} = 0$, which is true everywhere except at sources and sinks (charges), where it is ρ/ϵ and the analogy allows for this.

75. Maxwell (1855–6), p. 157.
76. Ibid., p. 208.
77. I have left Thomson (Lord Kelvin) out of this analysis because his 'field', like that of Maxwell, is a state of a Newtonian aether. He should be included in a more detailed analysis because of his influence on both Faraday and Maxwell, and because his conception of a 'vortex' aether was the most widely accepted at the turn of the century.
78. Maxwell (1855–6), p. 156.
79. Larmor (1937), May 15, 1855, p. 11. He also expressed the concern that Thomson had an analysis comparable to his own "lying around" somewhere.
80. Maxwell (1855–6), pp. 155–6.
81. Ibid., p. 156.
82. There has been much written on the subject of Maxwell's use of analogy. I will discuss some of that literature in what follows. In my view, the most useful discussion is that of North (1980). See, also, my (1984).
83. Maxwell (1855–6), p. 157.
84. Maxwell (1861–2), p. 564.
85. Ibid.
86. Ibid., 451.
87. Ibid., 452.
88. My position is, thus, in disagreement with that of Chalmers (1973) who maintains that the success of the analogy "as a heuristic guide was slight." The types of mistakes Maxwell made did not influence the role of the physical analogy: to act as a *guide* for exploring a new conceptual possibility. Also, as we will see, it is a consequence of his analogy that displacement produces magnetic effects – even if Maxwell did not follow this through at the time. But, I also disagree with Hesse (1973) who attempts to make more out of Maxwell's use of analogy than he, himself, did; i.e., she claims that physical analogy is a form of induction. I agree that his use of analogy is not hypothetico-deductive, since Maxwell explicitly states that his analogy in no way constitutes a physical hypothesis. However, if Maxwell could have deduced his equations from the experiments in the way she suggests, there would have been little need for the analogy in the first place. Maxwell saw the experiments as providing the basis for the construction of the initial analogy and for the alteration of it as his analysis progressed.
89. See Bromberg (1968b). Also, Maxwell said nothing about static electricity in his opening remarks. He mentioned only that he wanted to connect magnetism with electromagnetism and induced currents. He

expressed "surprise" at his final results in his letter of December, 1861 (Larmor (1937)).

90. Larmor (1937), letter of December, 1861. See also Maxwell (1861–2), p. 453.
91. Maxwell (1854).
92. Maxwell (1861–2), pp. 453–4.
93. A 'tensor', loosely characterized, is a generalization of the concept of a 'vector', which can be seen as a tensor of 'rank-1', i.e., as having one index. The 'stress tensor' is a rank-2 tensor. It is interesting that we still call this the 'stress' tensor even though without the medium, the term no longer has its original significance.
94. Maxwell (1861–2), p. 489.
95. It should be noted that the form of the equation is different from that which we use today. Maxwell's form is $\mathbf{E} = (\mathbf{v} \times \mathbf{B}) + \partial \mathbf{A}/\partial t - \nabla\Psi$, where '$\mathbf{v} \times \mathbf{B}$' is the effect of the motion of the body through the field, '$\mathbf{A}$' is the 'electrotonic state' and 'Ψ' is the electric tension at each point in space. Maxwell's $\mathbf{E}$ ($\mathbf{P}$, $\mathbf{Q}$, $\mathbf{R}$) here is the force-per-unit-charge on a charge moving with velocity $\mathbf{v}$. Our $\mathbf{E}$ is $(\mathbf{P}, \mathbf{Q}, \mathbf{R}) - (\mathbf{v} \times \mathbf{B})$, which gives us curl $\mathbf{E} = -\partial \mathbf{B}/\partial t$ which is customary.
96. Maxwell (1861–2), p. 483.
97. The '$(\mathbf{v} \times \mathbf{B})$' term corresponds to the 'cutting' by motion of the conductor through the field, while the '$\partial \mathbf{A}/\partial t$' corresponds to 'cutting' in virtue of the variation of the field itself (in Faraday's terms, "motion of the lines of magnetic force").
98. Duhem (1962), p. 98. See also, Chalmers (1973), p. 137.
99. For a detailed and penetrating analysis of Maxwell's derivation of the displacement current, see Bromberg (1968a and b).
100. Maxwell (1861–2), pp. 490–1.
101. Ibid., p. 500.
102. Duhem (1962), Bromberg (1968b), and Chalmers (1973). The equation in question is $\mathbf{R} = -4\pi E^2\mathbf{h}$, where '$\mathbf{R}$' is the electromotive force ($\mathbf{E}$); '$\mathbf{h}$', the displacement ($\mathbf{D}$), and 'E', a constant which depends upon the elasticity of the medium (c).
103. My interpretation is in agreement with the more detailed analysis of Bromberg (1968b).
104. That is, the 'stretching' of the particles would cause some motion of the magnetic vortices.
105. Maxwell (1854).
106. Duhem was the first to spot the mistake.
107. Maxwell and Jenkin (1863). I was made aware of this paper by Everitt (1974).
108. Larmor (1937), p. 35. See also, Maxwell's letter to Faraday in Campbell and Garnett (1969), p. xxii.
109. Maxwell (1861–2), p. 500.
110. Ibid., p. 486.
111. Maxwell (1864). He presents a more straightforward derivation in (1868).

112. Maxwell (1891), 2, p. 470.
113. Ibid., p. 470.
114. He did not "illustrate" his general mechanical analogy until Maxwell (1891), 2, p. 228. Everitt (1974) has pointed out that there is an "illustration" in the manuscript version of the paper, but this was deleted in the printed version. Maxwell proposed a general analogy between inductive circuits and coupled dynamical systems. He illustrated the latter by considering the action to be like that of two horses pulling on the swingle bar of a carriage.
115. Maxwell (1864), p. 563.
116. Maxwell (1890b), p. 317.
117. Maxwell (1890e), p. 781.
118. See, Maxwell (1876, 1877, 1890e, and 1891).
119. Maxwell (1864), p. 527.
120. Ibid.
121. Maxwell (1891), 2, p. 198.
122. Maxwell (1890e), pp. 783–4.
123. See Maxwell (1891), 2, p. 493.
124. Maxwell (1864), p. 529. He later offered two 'proofs' for the existence of the light aether. In Maxwell (1879c), he argued that light, itself, could not be a substance because of interference (pp. 764–5). In Maxwell (1877) he argued that the aether must exist because luminiferous energy is transmitted through empty space with a finite velocity (pp. 89–90).
125. Maxwell (1864), pp. 531–3.
126. Ibid., p. 564.
127. Here the sign for the displacement is positive. This creates a problem, not noticed by Maxwell, since Coulomb's law does not come out right. However, Maxwell had no reason to make the displacement negative, now that he had 'scrapped' the model. This further underscores the importance of that analogy as a guide to his thinking and again shows how he did not use it as a physical hypothesis.
128. Maxwell (1864), p. 577.
129. The 'wave equation' is $\nabla^2\mathbf{B} = k(\partial^2\mathbf{B}/\partial t^2)$. If this is correct, then electromagnetic actions propagate in waves, like light. The displacement current is the crucial factor in the derivation of this equation.
130. Maxwell (1864), p. 580. Also, in Maxwell (1891) he argued that it is not 'philosophical' to fill space with a new medium every time a new phenomenon is to be explained. However, if two different avenues lead to one medium, more evidence is provided for its existence.
131. Maxwell has one extra equation in the *Treatise* (1891), 2, p. 243. It represents the electromagnetic force on a conductor carrying an electric current through a magnetic field. Everitt (1974) claims that this is what we call the 'Lorentz force', in which case Maxwell would have to be credited with its derivation. It is not clear to me how to interpret this equation. However, the Lorentz force is a non-Newtonian force, which is not the case with Maxwell's.

132. Maxwell also wanted to give a field formulation for gravitational forces, but was "unable to conceive of the forces" in such a field (1864), p. 571. It should be noted that there is a formulation of electromagnetism in which electromagnetic actions are not instantaneous, yet the field is not essentially involved in the description of the phenomena, i.e., the Wheeler-Feynman formulation. It, however, has two objectionable components: (1) the use of 'advanced potential' solutions (what are called 'advanced' and 'retarded' potential solutions are both consequences of Maxwell's equations. However, the use of 'advanced' solutions is problematic because with them the effects of events precede the events themselves); and (2) the assumption that all radiation emitted is eventually absorbed. Also, this formulation is not in accord with Maxwell's conception.
133. Maxwell (1891), p. 493.

Chapter 5

134. Illy (1981), p. 184.
135. For excellent technical analyses see, Hirosige (1969) and Miller (1981).
136. "Maar men kan ook van goede dingen te veel hebben en zoo kan men ook door . . . al te aanschouwelijk te zijn, zijn doel voorbijstreven en hetgeen als beeld moet dienen zoo op den voorgrond stellen, dat het te veel voor de zaak zelf wordt gehouden . . . .

    Tegen eene dergelijke overmaat van aanschouwelijkheid nu heeft men zich vooral te wachten, als er in de natuurkunde van *krachten* sprake is." Lorentz (1878a), pp. 6–7.
137. "Even als hier is ook in andere gevallen het woord kracht slechts een naam voor sommige grootheden, die in onze wiskundige formules voorkomen." Ibid.
138. Klein (1957), p. xi.
139. De Haas-Lorentz (1957), p. 12.
140. Einstein (1949), p. 36.
141. He made this comment in a letter written in response to Poincaré's criticism that his aether violated the law of action and reaction. We will discuss this later in the chapter.
142. "Maar veroorlooft mij dan, U eraan te herinneren, dat zonder hypothesen nu eenmaal geene natuurverklaring denkbaar is. Het 'hypotheses non fingo' van Newton moet zeker cum grano salis worden opgevat en toen Ampère meende, dat hij de wetten der electrodynamica uit de ervaring alleen had afgeleid, vergat hij éène onderstelling, waarop zijn gansche gebouw berustte. Hypothesen zullen wij wel altijd blijven maken; wij hebben er slechts voor te waken, dat wij in dit gebruik onzer verbeeldingskracht niet al te veel behagen gaan scheppen." Lorentz (1891), p. 95.
143. Berkson (1974) was the first to note that Lorentz' stationary aether dealt a 'death blow' to the mechanical aether program.

144. von Helmholtz (1870).
145. Lorentz (1875).
146. "... dat aan de theorie de meest rechtstreeksche opvatting der feiten ten grondslag ligt." Ibid., p. 30.
147. Lorentz (1878b).
148. "Poincaré verhaalt van een natuurkundige, die verklaarde de geheele theorie van Maxwell verstaan te hebben, maar toch niet recht begrepen te hebben, wat nu een geelectriseerde bol was." Lorentz (1891), p. 95.
149. Lorentz (1878b), p. 24.
150. Lorentz (1892a).
151. Lorentz (1891).
152. Ibid., see esp. pp. 97–99.
153. "Wel niemand zal het tegenwoordig onbekend zijn, dat de natuurkundigen zich elk lichaam voorstellen als een stelsel van zeer kleine deeltjes, de zoogenaamde *moleculen*, waarvan elke, zooals vooral de scheikunde ons leert, uit een aantal nog kleinere deeltjes, de *atomen*, kan zijn opgebouwd." Lorentz (1878a), pp. 3–4.
154. "De opmerking schijnt mij niet zonder belang, dat men door eene kleine wijziging eene toenadering, ten minste wat den vorm betreft, van de nieuwe opvatting tot de oude kan bewerken. Men heeft vroeger op de deeltjes der denkbeeldige electrische stoffen overgedragen, wat men bij geladen geleiders had waargenomen. Iets dergelijks kan een volgeling van Maxwell doen. Men kan aannemen, dat er kleine electrisch geladen deeltjes bestaan, d.w.z. deeltjes met dergelijke eigenschappen als een geladen conductor, en onderstellen, dat eene voor ons waarneembare lading – ik bedoel de lading van een lichaam van waarneembare afmetingen – bestaat in eene opeenhooping van dergelijke deeltjes, en een electrische stroom in eene beweging daarvan." Lorentz (1891), p. 100.
155. Lorentz (1886).
156. For a good technical analysis of Lorentz' attempted 'mechanical' explanation, see McCormmach (1970).
157. "Freilich verstiesse diese Auffassung gegen den Satz von der Gleichheit der Wirkung und Gegenwirkung –, da wir ja Grund haben zu sagen, dass der Aether Kräfte auf die ponderable Materie *ausübe* –; aber, soviel ich sehe, zwingt nichts dazu, jenen Satz zu einem unbeschränkt gültigen Fundamentalgesetze zu erheben." Lorentz (1895), p. 28.
158. "Mais faut-il en vérité que nous nous en inquiétons." Lorentz to Poincaré, January 20, 1901. It should be noted that the *total* momentum of the system is conserved.
159. For a more detailed discussion, see, e.g., Whittaker (1951) and Schaffner (1972).
160. $1 - 1/n^2$, n = index of refraction.
161. Maxwell (1879c), p. 770.
162. For a detailed account of the history of the 'aether-drift' experiments, see Swenson (1973).
163. Lorentz (1886).

164. Lorentz followed-up his discussion of Stokes' aether hypothesis in (1892d) and (1902b). Berkson, quite correctly, has pointed out that, in retrospect, Lorentz' arguments are not as strong as he (Lorentz) had thought and that Stokes' hypothesis remains untested (Berkson, 1974, see esp., p. 327–329).
165. "Je désirais donc connaître les lois qui régissent les mouvements électriques dans des corps qui traversent l'éther sans l'entraîner . . . ." Lorentz (1892a), p. 168.
166. Lorentz' theory was the *only* one from which it was possible to deduce the Fresnel coefficient. In his (1895), he derived an even more precise expression. As Berkson has pointed out, these derivations were crucial in the rapid acceptance of Lorentz' theory by the physics community.
167. Lorentz (1895).
168. Lorentz to Rayleigh, August 18, 1892.
169. Lorentz (1892b), pp. 221–223.
170. This is in disagreement with Zahar (1973), who argues that Lorentz *derived* the contraction hypothesis from his "molecular force hypothesis." Lorentz, himself, says the introduction is *ad hoc.* I am in agreement with Miller (1974), who also disputes Zahar's contention.
171. It accounts for more in (1904c).
172. Lorentz (1895).
173. My discussion of these transformations is in a simplified form. Lorentz' transformations require three systems: a moving system, a stationary system, and a "fictitious" system. This latter system has all the properties of one at rest in the aether, but it requires alterations to the Galilean transformations. I found Miller's analysis (1981) a useful guide in understanding Lorentz' reasoning here.
174. In this case I have used the English translation rather than the Dutch original since some paragraphs do not occur in the original. Lorentz usually made his own 'translations' and, in this case, he made alterations to the original paper. The changes are more a matter of focus than of substance. Lorentz (1899a).
175. Ibid., p. 437.
176. Ibid., 440.
177. Ibid.
178. Ibid., p. 442.
179. Lorentz (1904c).
180. Hirosige (1969), p. 199. We find this interpretation in Whittaker (1951) as well.
181. Lorentz (1909), pp. 229–30. In a note added in the second edition (1915), Lorentz remarked: "If I had to write the last chapter now, I should certainly have given a more prominent place to Einstein's theory of relativity by which the theory of electromagnetic phenomena in moving systems gains a simplicity that I had not been able to attain. The chief cause of my failure was my clinging to the idea that the variable t only can be considered as the true time and that my local

time t' must be regarded as no more than an auxiliary mathematical quantity. In Einstein's theory, on the contrary, t' plays the same part as t;..." (p. 321). He went on to explain Einstein's conception in some detail, but does not say that he now agrees with it – only that it should be given more consideration than he had done in 1909.

182. "Zou het niet wel zoo eenvoudig zijn, ons een stilstaanden aether voor te stellen, ten opzichte waarvan de aarde draait en in welken de electromagnetische golven zich voortplanten zonder zich aan de bewegingen der aarde te storen?" Lorentz (1917), p. 262.
183. "Het is altijd bedenkelijk, een weg van onderzoek geheel af te sluiten en misschien is het, alles samen genomen, goed den aether nog een kans te gunnen. Wie weet of er niet een tijd komt, waarin bespiegelingen over zijne structuur, waarvan wij ons onthouden, vruchtbaar en doelmatig blijken." Ibid.
184. Lorentz (1922), p. 1.
185. Ibid.
186. Lorentz (1923a).
187. Lorentz (1901–1902).
188. It was not unreasonable for Lorentz to retain his belief in the aether until around 1915. There were still scientists, in particular in Germany and in England, who were attempting to work out the 'electromagnetic world-view' of which the 'world-aether' was a significant part. Even Ehrenfest, in 1913, said that only proof of the existence of light quanta would settle the issue.
189. Goldberg (1969) gives a similar interpretation.
190. "Dass von *absoluter* Ruhe des Aethers nicht die Rede sein kann, versteht sich wohl von selbst. Der Ausdruch würde sogar nicht einmal Sinn haben; wenn ich der Kürze wegen sage, der Aether ruhe, so ist damit nur gemeint, dass sich der eine Theil dieses Mediums nicht gegen den anderen verschiebe und dass alle wahrnehmbaren Bewegungen der Himmelskörper relative Bewegungen in Bezug auf den Aether seien." Lorentz (1895), p. 4.
191. "Schliesslich ist ihm nur noch soviel Substantialität geblieben, dass man durch ihn ein Koordinatensystem festlegen kann." Lorentz (1910), p. 210.
192. For a technical analysis of Lorentz' interpretation of the transformation rules, see Miller (1981).
193. "... dat men het essentieele van het relativiteitsbeginsel kan aanvaarden zonder met de oude opvattingen omtrent ruimte en tijd te breken." Lorentz (1915a).
194. Lorentz (1922), p. 221.
195. "Wir suchen ... vergeblich nach einem zureichenden Grunde dafür, warum eins dieser Systeme geeigneter sein sollte, bei Formulierung der Naturgesetze als Bezugssystem zu dienen, als das andere; wir fühlen uns vielmehr dazu gedrängt, die Gleichberechtigung beider Systeme zu postulieren." Lorentz to Einstein, January 1915. I am grateful to

Dr. A. J. Kox for calling this letter, which is not part of the collection at the Museum Boerhaave, to my attention and for providing me with a transcription of it.

196. ". . . die beiden Systeme sich in Bezug auf den Aether in verschiedener Weise bewegen." Ibid.
197. "Gehen Sie hier nicht etwas zu weit, indem Sie eine persönliche Auffassung als selbtsverständlich hinstellen?" Ibid.
198. "Ich muss indes gestehen, dass ich diese Bemerkung erst gemacht habe, *nachdem* ich die Hypothese gefunden hatte." Ibid.
199. "Wenn man die 'Verkürzung' aus den Gleichungen der Relativitätstheorie ableitet (was natürlich an und für sich ganz gerechtfertigt ist) und nichts weiter zur Erläuterung hinzufügt, so läuft man Gefahr, den Eindruck zu erwecken, dass es sich hier um 'scheinbare' Dinge, und nicht um eine wirkliche physikalische Erscheinung handle; wenigstens habe ich hin und wieder bei Vertretern der Relativitätstheorie Äusserungen gefunden, die von einer derartigen Auffassung zu zeugen scheinen. Dem gegenüber kann man bemerken, dass wenn wir eine 'Änderung' beobachten . . . diese 'Änderung' nach dem gewöhnlichen Sprachgebrauch (und warum sollten wir uns daran nicht halten?) eine physikalische Erscheinung darstellt. Die Verkürzung eines in Bezug auf K in Bewegung gesetzten Stabes ist genau so reell wie die Ausdehnung bei Temperaturerhöhung . . . ." Ibid.
200. "Wir sehen ganz klar sowohl das 'neben', 'hinter', und 'über', wie auch das 'nach' einander. Dabei scheint mir eine unverkennbare Verschiedenheit zwischen der Raum- und der Zeitvorstellung vorhanden zu [sein], eine Verschiedenheit die Sie wohl auch nicht ganz bei Seite schaffen können. Sie können nicht die Zeitkoordinate als völlig gleichwertig mit den Raumkoordinaten betrachten . . . Was die Zeit betrifft, so haben wir wie mich bedünkt eine vollkommen klare Vorstellung von aufeinanderfolgenden Augenblicken und auch von der 'Gleichzeitigkeit'." Ibid.
201. ". . . das eine oder das andere Bild mit einem Äther zu schmücken, oder soll ich sagen zu verunzieren . . . ." Ibid.
202. Shankland (1963), p. 57.

Chapter 6

203. Einstein (1905c). I will make use of the translation provided by Miller (1981) because it contains corrections Einstein made to the published original.
204. Einstein (1949).
205. Einstein (1973), p. 264.
206. See, e.g., Holton (1973), Miller (1981), and Pais (1982).
207. Einstein (1905a, b, c, and d), (1907a and b), and (1910).
208. Einstein (1949), p. 53.

209. Ibid.
210. Ibid.
211. Einstein (1894 or 1895).
212. Pais (1982), p. 132.
213. See, Einstein (1953).
214. Einstein (1949), pp. 25–27.
215. Ibid., p. 19.
216. Ibid., p. 37.
217. Ibid.
218. Ibid.
219. Ibid.
220. See Klein (1962) and (1979). There is some controversy in the historical literature over what, precisely, Planck did. I am only concerned with how Einstein interpreted him at the time.
221. Einstein (1949), p. 45.
222. Hume (1874) and Mach (1883). It is interesting to note that the late publication of Hume's *Treatise* made him somewhat of a contemporary of Einstein!
223. Einstein (1949), p. 45.
224. Ibid.
225. Ibid., p. 53.
226. See, "Methods of theoretical physics," in Einstein (1973).
227. Miller (1981), p. 392.
228. Ibid.
229. Ibid.
230. Einstein (1907), p. 413. In this article he discussed in detail the differences between his theory and that of Lorentz. He argued that in order to get relativity theory from Lorentz' (1904), all that is required is to replace "Ortzeit" with "Zeit". Technically, as I have said previously, he is right. However, the conceptual change this requires is not so "slight".
231. Einstein (1949), p. 53.
232. Just how much influence Mach had on Einstein's thinking is a subject of dispute. Einstein later rejected Mach's philosophical views. In an article in *Nature* (August 18, 1923, p. 253), Einstein is quoted as saying: "To the extent that Mach was a good mechanician he was a deplorable philosopher." However, he went on to say that "on the other point, namely, that concepts can change, I am in complete agreement with Mach." This point was most crucial for the development of relativity. Mach, himself, rejected the theory of relativity. The most 'Machian' element of Einstein's paper is his so-called 'operational definition' of 'space', 'time', and 'simultaneity'. It is interesting to note that according to Popper, Einstein later expressed the feeling that his presentation of the argument in this way was one of his greatest mistakes.
233. Hume (1874), p. 358.
234. Ibid., p. 368.
235. Ibid., p. 342.

236. See esp., Shankland (1963). For a somewhat different statement by Einstein, see Ogawa (1979).
237. Miller (1981), p. 394.
238. See, Miller (1980).
239. Miller (1981), p. 414.
240. Einstein (1905d).
241. See, e.g., Einstein (1907), (1909), and (1910).
242. Miller (1981), p. 414.
243. "Indessen lehrt ein genaueres Nachdenken, dass diese Leugnung des Äthers nicht notwendig durch das spezielle Relativitätsprinzip gefordert wird. Man kann die Existenz eines Äthers annehmen; nur muss man darauf verzichten, ihm einen bestimmten Bewegungszustand zuzuschreiben, d.h. man muss ihm durch Abstraktion das letzte mechanische Merkmal nehmen, welches ihm Lorentz noch gelassen hatte." Einstein (1920), p. 9.
244. "Das spezielle Relativitätsprinzip verbietet uns, den Äther als aus zeitlich verfolgbaren Teilchen bestehend anzunehmen, aber die Ätherhypothese an sich widerstreitet der speziellen Relativitätstheorie nicht. Nur muss man sich davor hüten, dem Äther einen Bewegungszustand zuzusprechen.

    Allerdings erscheint die Ätherhypothese vom Standpunkte der speziellen Relativitätstheorie zunächst als eine leere Hypothese. In den elektromagnetischen Feldgleichungen treten ausser den elektrischen Ladungsdichten nur die Feldstärken auf. Der Ablauf der elektromagnetischen Vorgänge im Vakuum scheint durch jenes innere Gesetz völlig bestimmt zu sein, unbeeinflusst durch andere physikalische Grössen. Die elektromagnetischen Felder erscheinen als letzte, nicht weiter zurückführbare Realitäten, und es erscheint zunächst überflüssig, ein homogenes, isotropes Äthermedium zu postulieren, als dessen Zustände jene Felder aufzufassen wären.

    Anderseits [sic!] lässt sich aber zugunsten der Ätherhypothese ein wichtiges Argument anführen. Den Äther leugnen, bedeutet letzten Endes annehmen, dass dem leeren Raume keinerlei physikalische Eigenschaften zukommen. Mit dieser Auffassung stehen die fundamentalen Tatsachen der Mechanik nicht im Einklang." Ibid., pp. 10–11.
245. The claim that no causal interactions can occur instantaneously follows from the relativity of simultaneity. Instantaneous actions would necessarily occur with space-like separation between cause and effect. This is forbidden for causal interactions because there would then be the possibility of a framework in which the effect preceded the cause. Causal interactions can only occur with time-like or zero (speed of light) separation.
246. The Maxwell 'stress-tensor' becomes generalized in the special theory and is interpreted as representing the 'flow' of electricity and magnetism through the field, rather than as representing force on, or stress in, a medium. It is now usually known as the 'impulse-energy tensor'.

247. "That which underlies phenomena; the permanent substratum of things; that which receives modifications and is itself not a mode; that in which accidents or attributes inhere." *Oxford English Dictionary.*
248. "Nur die Vorstellung eines Lichtäthers als des Trägers der elektrischen und magnetischen Kräfte passt nicht in die hier dargelegte Theorie hinein; elektromagnetische Felder erscheinen nämlich hier nicht als Zustände irgendeiner Materie, sondern als selbständig existierende Dinge, die der ponderabeln Materie gleichartig sind und mit ihr das Merkmal der Trägheit gemeinsam haben." Einstein (1907), p. 413.
249. How to attain (1) was not immediately clear because given the principle of inertia of energy, it was quite probable that the inert mass of a body depended upon the gravitational potential. With regard to (2), if we replace the Laplacian of the Newtonian potential by the D'Alembertian operator, the equation becomes Lorentz invariant. However: (1) the equivalence of inertial and gravitational mass rules this out, (2) the correct values for the precession of the perihelion of Mercury is not given, and (3) light rays are not bent.
250. An excellent account of this struggle is to be found in Sewell (1975).
251. A 'metric space' is one in which it is possible to say of pairs of neighboring points that they can be numerically compared.
252. It should be noted that there is difficulty in stating global conservation laws for the general theory. Also, although it is not possible to carry out localization of the energy and momentum in a generally covariant way, since the distribution of energy and momentum is different when described with respect to different frames, the total value is conserved in any particular frame with respect to which the events are described.
253. It is interesting to note that in the general theory, all phenomena are explained without having to make any definite assumptions about the nature of matter – just as Maxwell was able to do with electromagnetical phenomena without making any assumptions about the nature of charged matter.
254. Einstein (1973), pp. 261–262.
255. See, e.g., Einstein (1929).
256. "Zusammenfassend können wir sagen: Nach der allgemeinen Relativitätstheorie ist der Raum mit physikalischen Qualitäten ausgestattet; es existiert also in diesem Sinne ein Äther. Gemäss der allgemeinen Relativitätstheorie ist ein Raum ohne Äther undenkbar; denn in einem solchen gäbe es nicht nur keine Lichtfortpflanzung, sondern auch keine Existenzmöglichkeit von Massstäben und Uhren, also auch keine räumlich-zeitlichen Entfernungen im Sinne der Physik. Dieser Äther darf aber nicht mit der für ponderable Medien charakteristischen Eigenschaft ausgestattet gedacht werden, aus durch die Zeit verfolgbaren Teilchen zur bestehen; der Bewegungsbegriff darf auf ihn nicht angewendet werden." Einstein (1920), p. 15.
257. Maxwell (1890b), p. 322.

Chapter 7

1. A good illustration of this is provided by what happened in chemistry in the 18th century. First the new chemical concepts were constructed and only afterwards was there a change in the terminology.
2. See Gooding (1980a).
3. In work as yet unpublished, David Gooding attempts to formulate an account of the experimental/procedural dimension in the making of meaning. I would like to thank him for allowing me to 'steal' the title for Part III, "The Making of Meaning," from a paper he presented at a conference.
4. See note 21 of Part II.
5. There is an extensive philosophical literature on analogies in science. The most useful are Harré (1970), Hesse (1963), and North (1980). Only North addresses the role of analogies as "instruments of cognitive meaning" (his expression).
6. See, e.g., Osherson and Smith (1981), Rosch (1975), Rosch and Mervis (1975), and Smith and Median (1981).
7. For an interesting analysis of the problem of defining concepts, see Fodor et al. (1980). However, I disagree with Fodor's own conclusion (Fodor, 1981) that, since we cannot 'define' concepts, all lexical concepts are unstructured and innate. See also, Smith and Median (1981).
8. Marovcsik (1981) has made a similar suggestion for concepts in general. Although I reject the essentialism of his view, his notion of an 'Aitiational frame' has influenced my notion of a 'meaning schema'.
9. Shapere (1966) and (1982a).

Bibliography

Achinstein, P. (1973): "On meaning dependence," in *Theories and Observation in Science*, R. Grandy, ed. (Englewood Cliffs, N.J.: Prentice-Hall Inc.)

Agassi, J. (1971): *Faraday as a Natural Philosopher* (Chicago: University of Chicago Press)

Agassi, J. (1975a): "The nature of scientific problems and their roots in metaphysics," reprinted in *Science in Flux*, Boston Studies in the Philosophy of Science 28 (Dordrecht: Reidel), pp. 208–239

Agassi, J. (1975b): "The confusion between physics and metaphysics in the standard histories of science," reprinted in *Science in Flux*, Boston Studies in the Philosophy of Science 28 (Dordrecht: Reidel), pp. 270–281

Ayer, A. J., ed. (1959): *Logical Positivism* (Glencoe, Ill.: The Free Press)

Bence-Jones, H. (1870): *The Life and Letters of Faraday*, vols. 1 and 2 (London: Longmans, Green, and Co.)

Berkson, W. (1974): *Fields of Force: The Development of a World View from Faraday to Einstein* (New York: John Wiley & Sons)

Bork, A. (1966): "Physics just before Einstein," *Science* 152 (no. 3722): 597–603

Bromberg, J. (1968a): "Maxwell's electrostatics," *American Journal of Physics* 36: 142–151

Bromberg, J. (1968b): "Maxwell's displacement current and his theory of light," *Archive for History of Exact Sciences* 4: 218–234

Brush, S. G. (1967): "Notes on the history of the Fitzgerald-Lorentz contraction," *Isis* 58: 230–232

Campbell, L. and Garnett, W. (1969): *The Life of James Clerk Maxwell* (with a new preface and appendix with letters by Robert H. Kargon), reprint of 1882 edn. (London) with selections of letters from 2nd edn. (1884) (New York: Johnson Reprint Corp.)

Carnap, R. (1956a): "Empiricism, semantics and ontology," in *Meaning and Necessity* (Chicago: University of Chicago Press)

Carnap, R. (1956b): "The methodological character of theoretical concepts," in *The Foundations of Science and the Concepts of Psychology and Psychoanalysis*, H. Feigl and M. Scriven, eds., Minnesota Studies in the Philosophy of Science 1 (Minneapolis: University of Minnesota Press), pp. 38–76

Chalmers, A. F. (1973): "Maxwell's methodology and his application of it to electromagnetism," *Studies in the History and Philosophy of Science* 4(2): 107–164

Darden, L. (1980): "Theory construction in genetics," in *Scientific Discovery: Case Studies*, T. Nickles, ed., Boston Studies in the Philosophy of Science 60 (Dordrecht: Reidel), pp. 151–170

Darden, L. and Maull, N. (1977): "Interfield theories," *Philosophy of Science* 44: 43–64

Duhem, P. (1962): *The Aim and Structure of Physical Theory*, trans. P. P. Wiener (New York: Atheneum)

Ehrenfest, P. (1913): "Zur Krise der Lichtaether-hypothese," Inaugural address at Leiden University, reprinted in *Collected Scientific Papers*, M. J. Klein, ed., (Amsterdam: North Holland, 1957), pp. 306–327

Einstein, A. (1894 or 5): "Über die Untersuchung des Ätherzustandes im magnetischen Felde," unpublished manuscript printed in Mehra, J. (1971)

Einstein, A. (1905a): "Über einen Erzeugung und Verwandlung des Lichts betreffenden heuristischen Gesichtspunkt," *Annalen der Physik* 17: 132–148; trans. A. B. Arons and M. B. Peppard, *American Journal of Physics* 33: 367–374, 1965

Einstein, A. (1905b): "Die von der molekularkinetischen Theorie der Wärme geforderte Bewegung von in ruhenden Flüssigkeiten suspendierten Teilchen," *Annalen der Physik* 17: 549–560; reprinted in *Investigations on the Theory of Brownian Movement*, trans. A. D. Cowper, notes by R. Furth (New York: Dover, 1956)

Einstein, A. (1905c): "Zur Elektrodynamik bewegter Körper," *Annalen der Physik* 17: 891–921; trans. in Miller, A. (1981), pp. 391–415

Einstein, A. (1905d): "Ist die Trägheit eines Körpers von seinem Energieninhalt abhängig?" *Annalen der Physik* 18: 639–641

Einstein, A. (1907a): "Über die vom Relativitätsprinzip geforderte Trägheit der Energie," *Annalen der Physik* 23(4): 371–384

Einstein, A. (1907b): "Über das Relativitätsprinzip und die aus demselben gezogenen Folgerungen," *Jahrbuch Radioaktivität* 4: 411–462; 5: 98–99 Berichtigungen

Einstein, A. (1909): "Über die Entwicklung unserer Anschauungen über das Wesen und die Konstitution der Strahlung," *Physikalische Zeitschrift* 10: 817–826

Einstein, A. (1910): 'Principe de relativité et ses conséquences dans la physique moderne," *Archives des sciences physiques et naturelles* 29(4): 5–28, 125–144

Einstein, A. (1915): "Die Relativitätstheorie," in *Die Kultur der Gegenwart: Die Physik*, E. Lecher, ed. (Leipzig: Teubner), Teil 3, Abt. 3, pp. 703–713

Einstein, A. (1916a): "Grundlagen der allgemeinen Relativitätstheorie," *Annalen der Physik* 49: 769–822

Einstein, A. (1916b): "Ernst Mach," *Physikalische Zeitschrift* 17: 101–104

Einstein, A. (1917): *Relativity: The Special and General Theory*, trans. R. W. Lawson (New York: Crown Publishers, 1961)

Einstein, A. (1920): *Äther und Relativitätstheorie: Rede gehalten am 5. Mai 1920 an der Reichs-Universität zu Leiden* (Berlin: Springer)

Einstein, A. (1922): *Sidelights on Relativity*, trans. G. B. Jeffery and W. Perret (London: Methuen)

Einstein, A. (1923a): "Fundamental ideas and problems of the Theory of Relativity," reprinted in Nobel Lectures – Physics: 1901–1921 (New York: Elsevier, 1967), pp. 290–323

Einstein, A. (1923b): "Einstein and the philosophies of Kant and Mach," report in *Nature* 112: 253

Einstein, A. (1929): "Field theories, old and new," *New York Times*, Feb. 3

Einstein, A. (1930): "Raum, Äther und Feld in der Physik," *Forum Philosophicum* 1(2): 173–184

Einstein, A. (1931): "Maxwell's influence on the development of the conception of physical reality," in *Clerk Maxwell 1831–1931* (New York: Macmillan), pp. 66–73

Einstein, A. (1935): *The World as I See It*, trans. A. Harris (London: John Lane, The Bodley Lane)

Einstein, A. (1949): "Autobiographical notes," in Schilpp, P. A. (1970), pp. 2–94

Einstein, A. (1953): "The fundaments of theoretical physics," in *Readings in the Philosophy of Science*, H. Feigl and M. Brodbeck, eds. (New York: Appleton Century Crofts)

Einstein, A. (1973): *Ideas and Opinions* (New York: Dell)

Einstein, A. and Infeld, L. (1961): *The Evolution of Physics* (New York: Simon and Schuster)

Everitt, C. W. F. (1974): "Maxwell, James Clerk," in *The Dictionary of Scientific Biography*, vol. 9, Ch. C. Gillespie, ed. (New York: Charles Scribner's Sons), pp. 198–230

Faraday, M. (1821a): "Historical sketch of electromagnetism," *Annals of Philosophy* (new series) 2: 195–200, 274–290; 3: 107–121 (1822)

Faraday, M. (1821b): "On some new electro-magnetical motions and on the theory of magnetism," *Quarterly Journal of Science* 12

Faraday, M. (1831): *Experimental Researches in Electricity*, 3 vols (New York: Dover, 1965)

Faraday, M. (1844): "Matter," unpublished manuscript dated 19 February 1844, printed in Levere, T. H. (1968)

Faraday, M. (1859): *Experimental Researches in Chemistry and Physics* (London: Taylor and Francis)

Faraday, M. (1932): *Diary*, 7 vols., T. Martin, ed. (London: G. Bell & Sons Ltd.)

Feigl, H. (1970): "The 'orthodox' view of theories: Remarks in defense as well as critique," in *Analyses of Theories and Methods of Physics and Psychology*, M. Radner and S. Winokur, eds., Minnesota Studies in the Philosophy of Science 4 (Minneapolis: University of Minnesota Press)

Feyerabend, P. K. (1955): "Carnaps Theorie der Interpretation theoretischer Systeme," *Theoria* 21: 55–62

Feyerabend, P. K. (1958): "An attempt at a realistic interpretation of experience," *Proceedings of the Aristotelian Society* (new series) 58: 142–170.

Feyerabend, P. K. (1960): "Patterns of Discovery," *Philosophical Review* 59: 247–252

Feyerabend, P. K. (1962): "Explanation, reduction and empiricism," in *Scientific Explanation, Space, and Time*, H. Feigl and G. Maxwell, eds., Minnesota Studies in the Philosophy of Science 3 (Minneapolis: University of Minnesota Press), pp. 28–97

Feyerabend, P. K. (1965a): "Reply to criticism," in *Proceedings of the Boston Colloquium for the Philosophy of Science, 1962–1964*, R. S. Cohen and M. W. Wartofsky, eds., Boston Studies in the Philosophy of Science 2 (Dordrecht: Reidel)

Feyerabend, P. K. (1965b): "Problems of empiricism," in *Beyond the Edge of Certainty*, R. G. Colodny, ed. (Englewood Cliffs, N.J.: Prentice-Hall, Inc.), pp. 145–260

Feyerabend, P. K. (1969): "Problems of Empiricism II," in *Aim and Structure of Scientific Theory*, R. G. Colodny, ed. (Pittsburg: University of Pittsburg Press), pp. 275–353

Feyerabend, P. K. (1970a): "Classical empiricism," in *The Methodological Heritage of Newton*, R. E. Butts, ed. (Toronto: University of Toronto Press)

Feyerabend, P. K. (1970b): "Against method: outline of an anarchistic theory of knowledge," in *Analyses of Theories and Methods of Physics and Psychology*, M. Radner and S. Winokur, eds., Minnesota Studies in the Philosophy of Science 4 (Minneapolis: University of Minnesota Press), pp. 17–130

Feyerabend, P. K. (1972a): "Science without experience," in *Challenges to Empiricism*, H. Morick, ed. (Belmont, Calif.: Wadsworth), pp. 160–163

Feyerabend, P. K. (1972b): "How to be a good empiricist – a plea for tolerance in matters epistemological," in *Challenges to Empiricism*, H. Morick, ed. (Belmont, Calif.: Wadsworth), pp. 164–193

Feyerabend, P. K. (1973a): "On the interpretation of scientific theories," in *Theories and Observation in Science*, R. E. Grandy, ed. (Englewood Cliffs, N.J.: Prentice-Hall)

Feyerabend, P. K. (1973b): "On the 'meaning' of scientific terms," in *Theories and Observation in Science*, R. E. Grandy, ed. (Englewood Cliffs, N.J.: Prentice-Hall)

Fodor, J. A. (1981): *Representations: Philosophical Essays on the Foundations of Cognitive Science* (Brighton: The Harvester Press)

Fodor, J. A., Garrett, M. F., Walker, E. C. T., and Parkes, C. H. (1980): "Against definitions," *Cognition* 8: 263–367

Goldberg, S. (1969): "The Lorentz theory of electrons and Einstein's theory of relativity," *American Journal of Physics* 37: 982–994

Gooding, D. (1978): "Conceptual and experimental basis of Faraday's denial of electrostatic action at a distance," *Studies in the History and Philosophy of Science* 9: 117–149

Gooding, D. (1980a): "Faraday, Thomson, and the concept of magnetic field," *The British Journal for the History of Science* 13: 91–120

Gooding, D. (1980b): "Metaphysics versus measurement: The conversion and conservation of force in Faraday's physics," *Annals of Science* 37: 1–29

Gooding, D. (1981): "Final steps to the field theory: Faraday's study of magnetic phenomena, 1845–1850," *Historical Studies in the Physical Sciences* 11: 231–275

Gooding, D. (1982): "Empiricism in practice: Teleology, economy and observation in Faraday's physics," *Isis* 73: 46–67

Gruber, H. and Vonèche, J. J., eds. (1977): *The Essential Piaget* (New York: Basic Books, Inc.)

de Haas-Lorentz, G. L. (1957): *H. A. Lorentz: Impressions of His Life and Work* (Amsterdam: North Holland)

Hanson, N. R. (1961): *Patterns of Discovery* (Cambridge: Cambridge University Press)

Hanson, N. R. (1970): "A picture theory of meaning," in *Analyses of Theories and Methods of Physics and Psychology*, M. Radner and S. Winokur, eds., Minnesota Studies in the Philosophy of Science 4 (Minneapolis: University of Minnesota Press), pp. 131–141

Harré, R. (1970): *The Principles of Scientific Thinking* (London: Macmillan)

Heilbron, J. L. (1981): "The electrical field before Faraday," in *Conceptions of Ether,* G. N. Cantor and M. J. S. Hodge, eds. (Cambridge: Cambridge University Press), pp. 187–214

Heimann, P. M. (1971): "Faraday's theories of matter and electricity," *British Journal for the Philosophy of Science* 5 (no. 19): 235–257

Helmholtz, H. von (1870): "Über die Theorie der Elektrodynamik. Erste Abhandlung. Über die Bewegungsgleichungen der Elektricität für ruhende leitende Körper," *Journal für die reine und angewandte Mathematik* 72; reprinted in *Wissenschaftliche Abhandlungen von Hermann von Helmholtz*, vol. 1 (Leipzig, 1882)

Hempel, C. G. (1952): *Fundamentals of Concept Formation in Empirical Science*, vol. 2, no. 7, International Encyclopedia of Unified Science (Chicago: University of Chicago Press)

Hempel, C. G. (1958): "The theoretician's dilemma: A study in the logic of theory construction," in *Concepts, Theories, and the Mind-Body Problem*, H. Feigl, M. Scriven, and G. Maxwell, eds., Minnesota Studies in the Philosophy of Science 2 (Minneapolis: University of Minnesota Press), pp. 37–98

Hempel, C. G. (1970): "On the 'standard conception' of scientific theories," in *Analyses of Theories and Methods of Physics and Psychology*, M. Radner and S. Winokur, eds., Minnesota Studies in the Philosophy of Science 4 (Minneapolis: University of Minnesota Press), pp. 141–163

Hesse, M. (1962): *Forces and Fields: The Concept of Action at a Distance in the History of Physics* (Westport: Conn.: Greenwood Press)

Hesse, M. (1963): *Models and Analogies in Science* (London: Sheed and Ward)

Hesse, M. (1970): "An inductive logic of theories," in *Analyses of Theories and Methods of Physics and Psychology*, M. Radner and S. Winokur, eds., Minnesota Studies in the Philosophy of Science 4 (Minneapolis: University of Minnesota Press), pp. 164–180

Hesse, M. (1972): "Duhem, Quine and a new empiricism," in *Challenges to Empiricism*, H. Morick, ed. (Belmont, Calif.: Wadsworth), pp. 208–228

Hesse, M. (1973): "Logic of discovery in Maxwell's electromagnetic theory," in *Foundations of Scientific Method: The 19th Century*, R. N. Giere and R. S. Westfall, eds. (Bloomington: Indiana University Press), pp. 86–114

Hirosige, T. (1969): "Origins of Lorentz' theory of electrons and the concept of the electromagnetic field," *Historical Studies in the Physical Sciences* 1: 151–209

Holton, G. (1973): *Thematic Origins of Scientific Thought: Kepler to Einstein* (Cambridge, Mass.: Harvard University Press); Part II "On relativity theory"

Hume, D. (1874): *A Treatise on Human Understanding*, 2 vols. (London: Longman, Green and Co.)

Illy, J. (1981): "Revolutions in a revolution," *Studies in the History and Philosophy of Science* 12: 3, 173–210

Jackson, J. D. (1962): *Classical Electrodynamics* (New York: John Wiley)

Klein, M. J., ed. (1957): *Paul Ehrenfest: Collected Scientific Papers* (Amsterdam: North Holland)

Klein, M. J. (1962): "Max Planck and the beginnings of quantum theory," *Archive for History of Exact Sciences* 1: 459–479

Klein, M. J. (1973): "Mechanical explanation at the end of the nineteenth century," *Centaurus* 17: 58–79

Klein, M. J. (1979): "Einstein and the development of quantum physics," in *Einstein: A Centenary Volume*, A. P. French, ed. (London: Heinemann), pp. 133–152

Kripke, S. (1980): *Naming and Necessity* (Cambridge, Mass.: Harvard University Press), revised version of article with same name in *The Semantics of Natural Language*, D. Davidson and G. Harman, eds. (Dordrecht: D. Reidel, 1972)

Kuhn, T. S. (1962): *The Structure of Scientific Revolutions*, International Encyclopedia of Unified Science, vol. 2, no. 2 (Chicago: University of Chicago Press), 2nd edn. 1970

Kuhn, T. S. (1972): "Incommensurability and paradigms," in *Challenges to Empiricism,* H. Morick, ed. (Belmont, Calif.: Wadsworth), pp. 194–207

Lakatos, I. and Musgrave, A. (1970): *Criticism and the Growth of Knowledge* (Cambridge: Cambridge University Press)

Larmor, J., ed. (1937): *The Origins of Clerk Maxwell's Electric Ideas* (Cambridge: Cambridge University Press)

Levere, T. H. (1968): "Faraday, matter, and natural theology – Reflections on an unpublished manuscript," *British Journal for the History of Science* 4: 95–107

Lewis, C. I. (1952): "The given element in empirical knowledge," *Philosophical Review* 61

Lewis, C. I. (1956): *Mind and the World Order* (New York: Dover Publications)

Lorentz, H. A. (1875): "Over de theorie der terugkaatsing en breking van het licht," in *Collected Papers*, vol. 1 (The Hague: Martinus Nijhoff, 1935–1939), pp. 1–192

Lorentz, H. A. (1878a): "De moleculaire theorieën in de natuurkunde," in *Collected Papers*, vol. 9, pp. 1–25

Lorentz, H. A. (1878b): "Concerning the relation between the velocity of propagation of light and the density and composition of media," in *Collected Papers*, vol. 2, pp. 1–119

Lorentz, H. A. (1886): "Over den invloed dien de beweging der aarde op de lichtverschijnselen uitoefent," *Versl. Kon. Akad. Wetensch. Amsterdam* 2: 297–372; French translation, *Collected Papers*, vol. 4, pp. 153–214

Lorentz, H. A. (1891): "Electriciteit en aether," in *Collected Papers*, vol. 9, pp. 89–101

Lorentz, H. A. (1892a): "La théorie électromagnétique de Maxwell et son application aux corps mouvants," in *Collected Papers*, vol. 2, pp. 164–343

Lorentz, H. A. (1892b): "The relative motion of the earth and the ether," in *Collected Papers*, vol. 4, pp. 219–223

Lorentz, H. A. (1892c): "On the reflection of light by moving bodies," in *Collected Papers*, vol. 4, pp. 215–218

Lorentz, H. A. (1892d): "Stokes' theory of aberration," in *Collected Papers*, vol. 4, pp. 224–231

Lorentz, H. A. (1895): "Versuch einer Theorie der electrischen und optischen Erscheinungen in bewegten Körpern," in *Collected Papers*, vol. 5, pp. 1–137

Lorentz, H. A. (1897): "Concerning the problem of the dragging along of the ether by the earth," in *Collected Papers*, vol. 4, pp. 237–244

Lorentz, H. A. (1898): "Die Fragen, welche die translatorische Bewegung des Lichtäthers betreffen," in *Collected Papers*, vol. 7, pp. 101–115

Lorentz, H. A. (1899a): "Simplified theory of electrical and optical phenomena in moving systems," *Proc. Royal Acad. of Sci. Amsterdam* 1: 427–432

Lorentz, H. A. (1899b): "De aberratietheorie van Stokes in de onderstelling van een aether die niet overal dezelfde dichtheid heeft," *Versl. Kon. Akad. Wetensch. Amsterdam* 7: 523–529; French translation, *Collected Papers*, vol. 4, pp. 245–251

Lorentz, H. A. (1901–1902): "Aether theories and aether models," in *Lectures on Theoretical Physics*, vol. 1, H. Bremekamp, ed.;

trans. L. Silbersteen and A. P. H. Trivelli (London: Macmillan, 1927), pp. 3–71

Lorentz, H. A. (1902a): "Some considerations on the principles of dynamics, in connexion with Hertz's 'Prinzipien der Mechanik'," in *Collected Papers*, vol. 4, pp. 36–58

Lorentz, H. A. (1902b): "The fundamental equations for electromagnetic phenomena in ponderable bodies, deduced from the theory of electrons," in *Collected Papers*, vol. 3, pp. 117–131

Lorentz, H. A. (1902c): "The rotation of the plane of polarization in moving media," in *Collected Papers*, vol. 5, pp. 156–166

Lorentz, H. A. (1904a): "Maxwell's elektromagnetische Theorie," *Encykl. Math. Wiss.* 13: 63–144

Lorentz, H. A. (1904b): "Weiterbildung der Maxwellschen Theorie. Elektronentheorie," *Encykl. Math. Wiss.* 14: 145–228

Lorentz, H. A. (1904c): "Electromagnetic phenomena in a system moving with any velocity less than that of light," in *Collected Papers*, vol. 5, pp. 172–197

Lorentz, H. A. (1905): "De wegen der theoretische natuurkunde," in *Collected Papers*, vol. 9, pp. 53–76

Lorentz, H. A. (1909): *Theory of Electrons* (Leipzig: Teubner), 2nd edn. 1916; reprinted in 1952 (New York: Dover Publications)

Lorentz, H. A. (1910): "Alte und neue Fragen der Physik," in *Collected Papers*, vol. 7, pp. 205–257

Lorentz, H. A. (1915a): "De lichtaether en het relativiteitsbeginsel," in *Collected Papers*, vol. 9, pp. 233–243

Lorentz, H. A. (1915b): "Die Maxwellsche Theorie und die Elektronentheorie," *Die Kultur der Gegenwart: Die Physik*, E. Lecher, ed. (Leipzig: Teubner), pp. 311–333

Lorentz, H. A. (1917): "De gravitatietheorie van Einstein en de grondbegrippen der natuurkunde," in *Collected Papers*, vol. 9, pp. 244–263

Lorentz, H. A. (1921): "The Michelson-Morley experiment and the dimensions of moving bodies," in *Collected Papers*, vol. 5, pp. 356–362

Lorentz, H. A. (1922): *Problems of Modern Physics: Lectures at the California Institute of Technology* (Boston: Bateman, 1927); reprinted in 1967 (New York: Dover Publications)

Lorentz, H. A. (1923a): "The rotation of the earth and its influence on optical phenomena," in *Collected Papers*, vol. 7, pp. 173–178

Lorentz, H. A. (1923b): "Clerk Maxwell's electromagnetic theory," in *Collected Papers*, vol. 8, pp. 353–366

Mach, E. (1883): *Science of Mechanics*, trans. T. J. McCormack, 6th edn., revisions through 9th edn. (La Salle, Ill.: Open Court, 1960)

Mach, E. (1911): *History and Root of the Principle of the Conservation of Energy*, trans. Jourdain (La Salle, Ill.: Open Court)

Marovcsik, J. (1981): "How do words get their meanings?" *Journal of Philosophy* 88: 5–24

Maxwell, J. C. (1854): "On the equilibrium of elastic solids," in *The Scientific Papers of James Clerk Maxwell*, vol. 1, W. D. Niven, ed. (Cambridge: Cambridge University Press), pp. 30–74; reprinted in 1952 (New York: Dover Publications)

Maxwell, J. C. (1855–6): "On Faraday's lines of force," in *Scientific Papers,* vol. 1, pp. 155–229

Maxwell, J. C. (1856): "Are there real analogies in nature," reprinted in Campbell, L. and Garnett, W. (1969), pp. 235–244

Maxwell, J. C. (1861–2): "On Physical lines of force," in *Scientific Papers*, vol. 1, 451–513

Maxwell, J. C. (1864): "A dynamical theory of the electromagnetic field," in *Scientific Papers*, vol. 1, pp. 526–597

Maxwell, J. C. (1868): "On a method of making a direct comparison of electrostatic with electromagnetic force; with a note on the electromagnetic theory of light," in *Scientific Papers*, vol. 2, pp. 125–143

Maxwell, J. C. (1870): "Address to the mathematical and physical sections of the British Association," in *Scientific Papers*, vol. 2, pp. 215–229

Maxwell, J. C. (1872): "Review: Reprint of papers on electrostatics and magnetism, W. Thomson," in *Scientific Papers*, vol. 2, pp. 301–307

Maxwell, J. C. (1873): "Review: *Elements of Natural Philosophy*, W. Thomson and P. G. Tait," in *Scientific Papers*, vol. 2, pp. 324–328

Maxwell, J. C. (1876): "On the proof of the equations of motion of a connected system," in *Scientific Papers*, vol. 2, pp. 308–309

Maxwell, J. C. (1877): *Matter and Motion*, reprinted with appendices and notes by Sir J. Larmor, 1887 (New York: Dover Publications)

Maxwell, J. C. (1879a): "Attraction," in *Scientific Papers*, vol. 2, pp. 485–491

Maxwell, J. C. (1879b): "Constitution of bodies," in *Scientific Papers*, vol. 2, pp. 616–624

Maxwell, J. C. (1879c): "Ether," in *Scientific Papers*, vol. 2, pp. 763–773

Maxwell, J. C. (1890a): "On the mathematical classification of physical quantities," in *Scientific Papers*, vol. 2, pp. 257–266

Maxwell, J. C. (1890b): "On action at a distance," in *Scientific Papers*, vol. 2, pp. 311–323

Maxwell, J. C. (1890c): "Faraday," in *Scientific Papers*, vol. 2, pp. 353–360

Maxwell, J. C. (1890d): "Molecules," in *Scientific Papers*, vol. 2, pp. 361–378

Maxwell, J. C. (1890e): "Thomson and Tait's natural philosophy," in *Scientific Papers*, vol. 2, 776–785

Maxwell, J. C. (1891): *A Treatise on Electricity and Magnetism*, 3rd edn. (Oxford: Clarendon): reprinted in 1954 (New York: Dover Publications)

Maxwell, J. C. and Jenkin, F. (1863): "On the elementary relations of electrical qualities," *Report of the British Association for the Advancement of Science*, 1st series 32: 130–163

McCormmach, R. (1970a): "H. A. Lorentz and the electromagnetic view of nature," *Isis* 61: 459–497

McCormmach, R. (1970b): "Einstein, Lorentz, and the electron theory," *Historical Studies in the Physical Sciences* 2: 41–87

Mehra, J. (1971): "Albert Einsteins erste wissenschaftliche Arbeit," *Physikalische Blätter* 27: 386–391

Miller, A. I. (1974): "On Lorentz' methodology," *British Journal for the Philosophy of Science* 25: 29–45

Miller, A. I. (1980): "On some other approaches to electrodynamics in 1905," in *Some Strangeness in the Proportion*, H. Woolf, ed. (Reading, Mass.: Addison-Wesley)

Miller, A. I. (1981): *Albert Einstein's Special Theory of Relativity* (Reading, Mass.: Addison-Wesley)

Mischel, T., ed. (1971): *Cognitive Development and Epistemology* (New York: Academic Press)

Mayer, D. (1978): "Continuum mechanics and field theory: Thomson and Maxwell," *Studies in the History and Philosophy of Science* 9: 35–50

Nagel, E. (1971): "Theory and observation," in *Observation and Theory*, E. Nagel, S. Bramberger, and A. Grünbaum, eds. (Baltimore: The Johns Hopkins University Press)

Nersessian, N. J. (1979): "The roots of epistemological 'anarchy'," *Inquiry* 22: 423–440

Nersessian, N. J. (1982): "Why is 'incommensurability' a problem?" *Acta Biotheoretica* 31: 205–218

Nersessian, N. J. (1984): "Aether/or: The creation of scientific concepts," *Studies in the History and Philosophy of Science* 15 (no. 3)

Newell, A. and Simon, H. A. (1972): *Human Problem Solving* (Englewood Cliffs, N.J.: Prentice Hall)

North, J. D. (1980): "Science and analogy," in *On Scientific Discovery*, M. S. Grmek, R. S. Cohen, and G. Cimenò, eds. (Dordrecht: Reidel), pp. 115–140

Osherson, D. N. and Smith, E. E. (1981): "On the adequacy of prototype theory as a theory of concepts," *Cognition* 9: 35–58

Ogawa, T. (1979): "Japanese evidence for Einstein's knowledge of the Michelson-Morley experiment," *Japanese Studies in the History of Science* 18: 73–81

Pais, A. (1982): *'Subtle is the Lord . . . ': The Science and the Life of Albert Einstein* (Oxford: Oxford University Press)

Pauli, W. (1921): *Theory of Relativity*, trans. G. Field (New York: Pergamon, 1958)

Pearce, D. (1982): "Stegmüller on Kuhn and incommensurability," *British Journal for the Philosophy of Science* 33: 389–396

Poincaré, H. (1900): "La théorie de Lorentz et le principe de réaction," in *Recueil de Travaux Offerts par les Auteurs à H. A. Lorentz* (The Hague: Martinus Nijhoff), pp. 252–278

Popper, K. R. (1959): *The Logic of Scientific Discovery* (London: Hutchinson)

Popper, K. R. (1962): *Conjectures and Refutations* (New York: Basic Books)

Popper, K. R. (1983): *Realism and the Aim of Science*, W. W. Bartley III, ed., from the *Postscript to the Logic of Scientific Discovery* (London: Hutchinson)

Putnam, H. (1973): "Explanation and reference," in *Conceptual Change*, G. Pearce and P. Maynard, eds. (Dordrecht: Reidel), pp. 199–221

Putnam, H. (1975): "The meaning of meaning," in *Philosophical Papers*, vol. 2 (Cambridge: Cambridge University Press), pp. 215–271

Quine, W. V. O. (1960): *Word and Object* (Cambridge, Mass.: M.I.T. Press)

Quine, W. V. O. (1966): "On Carnap's views on ontology," in *The Ways of Paradox* (New York: Random House)

Quine, W. V. O. (1969): *Ontological Relativity and Other Essays* (New York: Columbia University Press)

Quine, W. V. O. (1970): "Grades of theoreticity," in *Experience and Theory*, Foster and Swanson, eds. (Amherst: University of Massachusetts Press)

Quine, W. V. O. (1973): *The Roots of Reference: The Paul Carus Lectures* (La Salle, Ill.: Open Court)

Rosch, E. (1975): "Cognitive representations of semantic categories," *Journal of Experimental Psychology: General* 104: 192–233

Rosch, E. and Mervis, C. B. (1975): "Family resemblance studies in the internal structure of categories," *Cognitive Psychology* 7: 573–605

Rosch, E. and Lloyd, B. B., eds. (1978): *Cognition and Categorization* (Hillsdale, N.J.: Erlbaum)

Schaffner, K. F. (1969): "The Lorentz electron theory of relativity," *American Journal of Physics* 37: 498–513

Schaffner, K. J. (1972): *Nineteenth-Century Aether Theories* (New York: Pergamon)

Scheffler, I. (1967): *Science and Subjectivity* (Indianapolis: Bobbs-Merrill)

Schilpp, P. A. (1970): *Albert Einstein: Philosopher-Scientist*, 3rd edn. (La Salle, Ill.: Open Court)

Sewell, W. (1975): *Einstein, Mach and the General Theory of Relativity*, unpublished Ph.D thesis, CWRU

Shankland, R. (1963): "Conversations with Albert Einstein", *American Journal of Physics* 31: 47–57

Shapere, D. (1966): "Meaning and scientific change," in *Mind and Cosmos*, R. G. Colodny, ed. (Pittsburg: University of Pittsburg Press), pp. 41–85

Shapere, D. (1974): "Scientific theories and their domains," in *The Structure of Scientific Theories*, F. Suppe, ed. (Urbana: University of Illinois Press), pp. 518–565

Shapere, D. (1980): "The character of scientific change," in *Scientific Discovery, Logic, and Rationality*, T. Nickles, ed. (Dordrecht: Reidel), pp. 61–116

Shapere, D. (1982a): "Reason, reference, and the quest for knowledge," *Philosophy of Science* 49: 1–23

Shapere, D. (1982b): "The concept of observation in science and philosophy," *Philosophy of Science* 49: 485–525

Siegel, D. (1981): "Thomson, Maxwell and the universal ether in Victorian physics," in *Conceptions of Ether*, G. N. Cantor and M. J. S. Hodge, eds. (Cambridge: Cambridge University Press), pp. 239–268

Simpson, T. K. (1970): "Some observations on Maxwell's *Treatise on Electricity and Magnetism*," *Studies in the History and Philosophy of Science* 1: 249–263

Smith, E. E. and Median, D. L. (1981): *Concepts and Categories* (Cambridge, Mass.: Harvard University Press)

Sneed, J. D. (1971): *The Logical Structure of Mathematical Physics* (Dordrecht: Reidel), 2nd edn., 1979

Spencer, J. B. (1967): "Boscovich's theory and its relation to Faraday's researches: An analytical approach," *Archive for History of Exact Sciences* 4: 184–202

Stegmüller, W. (1975): "Structures and dynamics of theories: Some reflections on J. D. Sneed and T. S. Kuhn," in *Collected Papers on Epistemology, Philosophy of Science and History of Philosophy*, vol. 2 (Dordrecht: Reidel), pp. 177–202

Stegmüller, W. (1976): *The Structure and Dynamics of Theories* (New York: Springer-Verlag)

Stein, H. (1967): "Newtonian space-time," *The Texas Quarterly*

Stein, H. (unpublished): "On action at a distance: Metaphysics and method in Newton and Maxwell"

Stein, H. (1970): "On the notion of field in Newton, Maxwell, and beyond," in *Historical and Philosophical Perspectives of Science*, R. H. Stuewer, ed., Minnesota Studies in the Philosophy of

Science 5 (Minneapolis: University of Minnesota Press), pp. 264–287

Stein, H. (1981): "'Subtler forms of matter' in the period following Maxwell," in *Conceptions of Ether*, G. N. Cantor and M. J. S. Hodge, eds. (Cambridge: Cambridge University Press), pp. 309–340

Stein, H. (unpublished): "Poincaré on hypothesis, electrodynamics, and relativity"

Suppe, F. (1977): *The Structure of Scientific Theories* (Chicago: University of Illinois Press), 2nd rev. edn.

Swenson, L. S. Jr. (1973): *The Ethereal Aether* (Austin: University of Texas Press)

Thomson, J. J. (1908): "On the light thrown by recent investigations on electricity in the relation between matter and ether," *Annual Report of the Smithsonian Institution*, pp. 233–244

Turner, J. (1955): "Maxwell on the method of physical analogy," *British Journal for the Philosophy of Science* 6: 226–238

Whittaker, E. (1951): *A History of the Theories of Aether and Electricity*, vols. 1 and 2 (New York: Philosophical Library)

Williams, L. P. (1964): *Michael Faraday: A Biography* (New York: Basic Books)

Williams, L. P., ed. (1971): *The Selected Correspondence of Michael Faraday*, 2 vols (Cambridge: Cambridge University Press)

Williams, L. P. (1975): "Review: Agassi's *Faraday as a Natural Philosopher* and Berkson's *Fields of Force*," *British Journal for the Philosophy of Science* 26: 241–253; response by Agassi, Berkson, and Williams, *British Journal for the Philosophy of Science* 29: 243–252

Wise, M. N. (1982): "The Maxwell literature and British dynamical theory," *Historical Studies in the Physical Sciences* 13: 175–205

Wittgenstein, L. (1968): *Philosophical Investigations*, 3rd edn., trans. G. E. M. Anscombe (New York: Macmillan)

Woodruff, A. E. (1968): "The contributions of Hermann von Helmholtz to electrodynamics," *Isis* 59: 300–311

Zahar, E. (1973): "Why did Einstein's programme supersede Lorentz's?" *British Journal for the Philosophy of Science* 24: 95–123, 223–262

Index